Classic Engines, Modern Fuel

Modern

Fuel

The Problems, the Solutions

Also available from Veloce Publishing:

1275cc A-Series High Performance Manual, The (Hammill)

Anatomy of the Classic Mini (Huthert & Ely)

Anatomy of the Works Minis (Moylan)

Build & Power Tune Weber & Dellorto DCOE, DCO/SP & DHLA Carburettors 3rd Edition, How To (Hammill)

BRM – A Mechanic's Tale (Salmon)

Classic British Car Electrical Systems (Astley)

Competition Car Aerodynamics 3rd Edition (McBeath)

Competition Car Composites A Practical Handbook (Revised 2nd Edition) (McBeath)

Ford Cleveland 335-Series V8 engine 1970 to 1982 (Hammill)

Immortal Austin Seven (Morgan)

Inside the Rolls-Royce & Bentley Styling Department – 1971 to 2001 (Hull)

Making a Morgan (Hensing)

MG Midget & Austin-Healey Sprite High Performance Manual, The (Stapleton)

Modify your retro or classic car for high performance, How to (Stapleton)

Morris Minor, 70 Years on the Road (Newell)

Power Tune Ford SOHC Pinto & Sierra Cosworth DOHC Engines, How to (Hammill)

Rover V8 – the story of the engine (Taylor)

SU Carburettor High Performance Manual, The (Hammill)

You & Your Jaguar XK8/XKR – Buying, Enjoying, Maintaining, Modifying – New Edition (Thorley)

Which Oil? – Choosing the right oils & greases for your antique, vintage, veteran, classic or collector car (Michell)

Works MGs, The (Allison & Browning)

Works Minis, The Last (Purves & Brenchley)

Works Rally Mechanic (Moylan)

www.veloce.co.uk

First published in February 2020 by Veloce Publishing Limited, Veloce House, Parkway Farm Business Park, Middle Farm Way, Poundbury, Dorchester DT1 3AR, England. Tel +44 (0)1305 260068 / Fax 01305 250479 / e-mail info@veloce.co.uk / web www.veloce.co.uk or www.velocebooks.com.
ISBN: 978-1-787115-90-3; UPC: 6-36847-01590-9.

Classic
Engines,
Modern
Fuel

The Problems, the Solutions

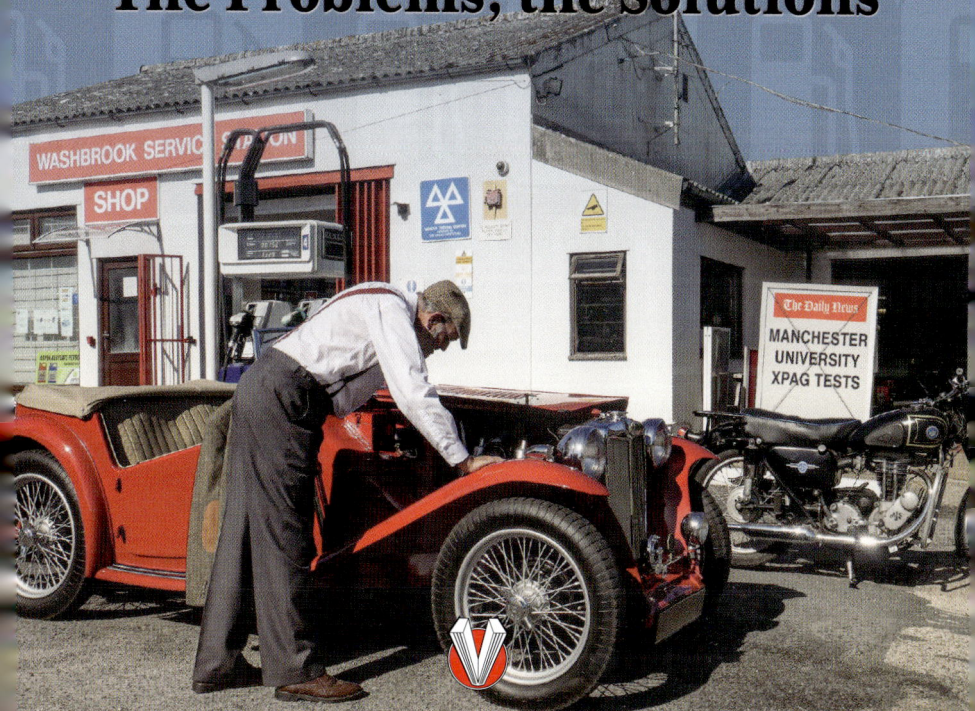

VELOCE PUBLISHING
THE PUBLISHER OF FINE AUTOMOTIVE BOOKS

Contents

Biography

Paul was born in the early 1950s and bought his first car, a 1949 MG TC, for £60 in 1967. Even though it had a valid MOT, it was in very poor condition. Before he could use it, Paul and his father rebuilt the TC. Removing the body, replacing the rotten wood, and giving it a brush coat of paint improved its looks. Refurbishing the suspension, wheels, steering and brakes made the car roadworthy. Finally, rewiring it with a home-made loom removed the risk of electrical fires. By then, Paul had no money left to rebuild the worn out engine.

Paul Ireland.

Even though it rattled and burned oil, Paul used his MG TC while studying at Manchester University. In the early 1970s, student cars were a rarity. Owning a two seater, MG sports car, no matter what condition, was a real status symbol.

Although he has a PhD in Experimental Nuclear Physics, Paul is the black sheep of the family; both his two sons, and his father are 'proper' engineers. He has inherited his family's practical abilities and a love for all things mechanical, always maintaining his MG car himself.

After the removal of lead from petrol in the late 1980s, Paul's MG did not

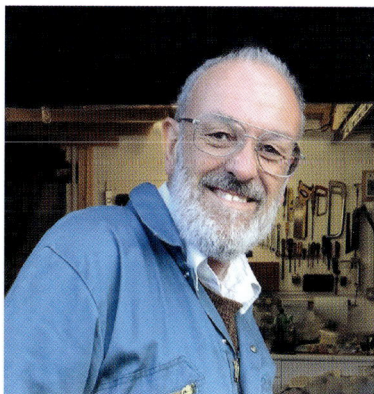

MG when bought.

run as well on modern petrol. Paul's training as a physicist has enabled him to research these problems, identifying and trialling solutions using his MG as a test bed. The practical application of this research is beneficial to all classic car owners.

In 2003, when he returned from university, Paul's oldest son stripped the MG back to the chassis. Now having more money, Paul completely restored his beloved TC. Winning prizes at prestigious shows, Paul now uses his MG for tours and longer trips. Needless to say, modern petrol is no longer the cause of any running problems or worries about damage to the engine.

Passo del Bernina.

Introduction

Petrol, or gasoline as it is called in some parts of the world, is that volatile liquid we use to fill our cars, motorcycles or powered tools. Mostly, we take it for granted, and often our only interest is finding the cheapest place to buy it. With that in mind, why should somebody go to the trouble of writing a book about it? And who would be interested in reading it?

Over the years, the makeup of petrol has significantly changed. As a result, some petrol (gas) engines do not run as well as they could and may be damaged when using modern petrol. This is particularly true of classic or old-timer cars and motorbikes. Owners suffer from a range of petrol-related problems. Widely-held myths claim these problems are the result of the removal of tetraethyl lead. Others suggest modern petrol burns hotter and slower than classic petrol. None of these myths are true.

One common problem is often referred to as the 'Hot Restart Problem'. This is the tip of the iceberg. Overheating and reports from some owners their vehicles run very rough after being stored for a few weeks are all problems caused by modern petrol.

My 1948 MG TC also suffered from these problems. After many years of trying to get it to run properly, I was having little success. I realised I would need to understand the root causes if I was to develop solutions that would work.

The Department of Mechanical, Aerospace and Civil Engineering (MACE) at Manchester University agreed to run a student project to investigate these problems. An engine, designed in the late 1930s, was installed in a test cell and used to investigate the way it ran on modern petrol.

This book goes beyond publishing the results of these investigations. Based on the underlying research, it reveals the fascinating truth about how a petrol or gas engine works and shows how modern petrol differs from classic petrol. It suggests ways to make an engine run better; and in the process it debunks some of the ideas about improving an engine's performance. Using the findings from this research, it advises on the best grades of petrol to use.

A more recent change to the composition of petrol is the addition of ethanol (alcohol). Referred to as 'ethanol blended petrol,' many suggest this will damage a vehicle's fuel system – a worry for all car owners. However, all is not what it appears. This book highlights the good, the bad and the ugly faces of ethanol blended petrol.

One surprising finding emerged from the tests. The potential for engine damage is at its worse when being driven at low to medium revolutions per minute (rpm) on part throttle, which is typical when driving on a public highway. In contrast, driving at high rpm on full throttle – for example, when racing – causes fewer problems. What is interesting is that race tuning an engine can make matters worse for driving on the public road.

The aim of this book is: to help the reader appreciate the complex inner workings of petrol (gasoline) engines, or, to give them their correct name, internal combustion, spark-ignition engines; to help readers understand the characteristics of modern petrol and why it causes the problems it does; and finally, to outline a set of practical steps owners can take to mitigate these problems, many of which have been tested and proven by the classic MG car fraternity. It will be of interest to anybody who owns a classic (old-timer) motor car or motorcycle, or who is a 'petrolhead'.

This book has four main sections, and readers can share comments and experiences using the online bulletin board (classicenginesmodernfuel.org.uk).

Chapters 2 and 3 describe the characteristics of modern and ethanol blended petrol;

Chapters 4 to 6 give an insight into the inner workings of a petrol engine and describes how a variable jet carburettor works;

Chapters 7 and 8 present the findings of the tests at Manchester. It gives the reason why modern petrol appears to burn slower and hotter than classic petrol, and warns of how this may cause damage to the engine;

Chapters 9 to 13 suggest the solutions: tuning the ignition timing and carburation to better suit modern petrol, ranking the performance of different grades of petrol, and suggesting ways to mitigate the Hot Restart Problem. The website https://classicenginesmodernfuel.org.uk/bestfuel/ allows owners to share information on brands and grades of fuel.

The last chapters suggest a simple means of assessing the efficiency of a water-cooled engine. This can be used to assess how well the engine runs on different grades of petrol (Chapter 14), and summarises the problems and suggested solutions (Chapter 15).

Finally, the appendix describes, in a simple and practical way, how to rebuild and tune SU carburettors. The advice is also applicable to Stromberg carburettors.

Acknowledgements

I would particularly like to thank the people who helped to run the tests at Manchester; David Houghton, Prof John Yates, Stuart Ray, and Peter Cole. The student project was organised by Dr Rob Prosser. Thank you to the students: Abdur Abdulrahman, Ahmed Oyeleke, Chin Lim, Maria Yankova, Norshah Shuaib, and Victor Okwuosa who rebuilt the engine and installed it in the test-cell.

Support from the following people and organisations made the tests possible: 123ignition-conversions, Andrew Owst, Anglo American Oil Company, BP Australia, Burlen Fuel Systems Ltd, Distributor Doctor, Driven Racing Oil, Federation of British Historic Vehicle Clubs, Innovate Motorsports, MG Car Club, MG T Register, MG Y Register, NTG Motor Services, Octagon Car Club, and Totally T Type 2.

I wish to acknowledge the department of Mechanical, Aerospace and Civil Engineering at The University of Manchester for allowing us to use their facilities to test the XPAG engine.

Thanks to Anders Hildebrand, Barrie Jones, David Heath, Garry Whitefield, George Wilder, John Burnett, John James, Lake Speed, Matt Vincent, Michael Harvey, Nigel Crowther and Nigel Stevens for their invaluable input.

Finally, I would also like to mention Vernon Byrom, the former editor of the *Octagon Car Club Magazine* for his support in publishing my early thoughts on the problems caused by modern petrol.

I dedicate this book to my wife Christine Ireland for her support, patience and understanding every time I emerged from the garage accompanied by a strong smell of petrol.

All royalties from this book will be given to help educate children in the developing world.

1 The XPAG tests

The tests run at Manchester University investigated the way an engine, designed in the 1930s, ran on different fuels. The aim was to identify the causes of problems many classic or old-timer car owners are experiencing when using modern petrol.

The tests are the most comprehensive ever to be performed on a classic engine using modern fuel. They answer many questions about the problems modern petrol causes. This chapter describes the background behind these tests and how they were run. The findings are the basis of this book and the practical recommendations to mitigate the problems caused by modern petrol.

Background

In the UK, those who can remember the 1970s, will know there was two-star and four-star leaded petrol. I used to run my MG TC on two-star. It ran like a dream. After the removal of lead from petrol in the late 1980s, my TC started to suffer from running problems, and I soon realised many classic car owners were experiencing similar issues. The cause: modern petrol. Unfortunately, there is very little published about the differences between classic and modern petrol. During my research, I contacted fuel experts who kindly provided advice. I also started testing different concoctions and types of petrol in my MG TC, writing articles about my findings in club magazines.

The problems caused by modern petrol are most obvious in classic vehicles.

Drive any further than 10 miles, stop for 10 minutes, for example to fill up with petrol, the engine will not re-start. This is referred to as the 'Hot Restart Problem.'

A related problem occurs in traffic. In years gone by, when driving in slow moving or stop-start traffic, the temperature gauge would creep up. As it reached 100°C (212°F) the cooling water would start to boil. The engine would misfire, finally coughing and spluttering to a stop. With modern petrol this can happen before the temperature gauge reaches 100°C (212°F). While the symptoms are the same as classic overheating, they occur when the temperature is within its normal running range. Annoying when you cannot start your car in a petrol station, dangerous if it stops in busy traffic.

These are the visible symptoms of more serious problems. Some owners have reported burned exhaust valves, exhaust ports or pistons, and cracked cylinder heads. There is a popular view that the reason for this is that modern petrol burns slower and hotter than classic petrol; this is not true. Burn rate and exhaust temperatures are no different. I have called this the 'Slow Combustion Problem.' This is a symptom of a more complex phenomenon suffered by all spark-ignition engines, called 'cyclic variability.' Understanding its cause makes it possible to put in place solutions to mitigate its effect and the serious damage it can cause.

My attempts to get my TC running properly started with rebuilding the ignition system and carburettors, and tuning my car. When this was not successful I took it on a rolling road, where the ignition timing was significantly advanced. This made the car run better, which appeared to support the popular view that modern petrol burned more slowly. I started using different brands, grades and mixtures of petrol in an attempt to improve matters even further. After discussing my results with fuel experts, to my surprise, I discovered my assumptions about burn rate were wrong.

Two things became obvious from these road tests. Firstly, modern petrol was not what it appeared. Different brands, grades and even the time of year it was purchased affected how well my TC ran. Secondly, the way an internal combustion engine worked was far more complex than I had thought.

I realised the only way to understand these differences was to run a set of tests in a controlled environment. In practice, this meant working with a university that had an engine test cell. This allows the engine to be instrumented and repeatedly run in the same way using different fuels. The School of Mechanical, Aerospace and Civil Engineering (MACE) at Manchester University had such facilities. They agreed to run a student project to investigate these problems.

Initially, the student project went well. I was able to borrow an XPAG engine, like the one fitted in my MG TC, and the students (Figure 1-0) installed this into the test cell and began their research. Unfortunately, they ran out of time before they were able to complete the tests.

With support from the MG Car Club, MACE, the Federation of British Historic Vehicle Clubs (FBHVC), David Houghton (who came out of retirement to manage the test cell), Prof John Yates, Stuart Ray and Peter Cole, I managed to complete these tests.

The analysis of this data has

1-0: The students.

identified the causes of the Hot Restart and Slow Combustion Problems. With a better understanding of these causes, it is possible to suggest solutions to mitigate their severity. These are discussed in Chapters 9 to 13.

Why use an XPAG engine?

While seeking funding for this research, people asked: "Why test an XPAG? They're old engines, designed in the late 1930s and only fitted to MG T Types? These problems affect most classic vehicles and other types of petrol engines."

1-1: David, myself and Stuart in the test cell.

The XPAG or 'X' series engines were used in nearly all Morris and Wolseley cars until 1956, including many thousands of Morris 10/4 utility cars and vans made during WW2. The XPAG is a good compromise. Its long stroke bottom end shares a great deal with engines dating from the 1920s and 1930s. The cylinder head design is virtually identical to the A and B series engines fitted to later BMC and British Leyland cars. The XPAG shares many of the characteristics of engines from different manufacturers used in most cars up to the 1990s.

XPAG engines		
	Cylinders	Straight-four in-line water-cooled, cast iron block
	Capacity	1250cc (76.28in^3)
	Bore × Stroke	66.5 × 90mm 2.62 × 3.54in
	Bore/stroke ratio	0.74 (long stroke)
	Compression ratio	7.25:1
	Crankshaft bearings	Three main bearings White metal in steel shells
	Valve gear	Overhead valve (OHV) Two valves per cylinder
	Power	56PS (55bhp) (41kW) at 5200rpm
	Torque	87Nm (64lb/ft) (8.9kgm) at 2700rpm
	Fuel system	Two SU carburettors

1-1a XPAG engine specification.

Would it be better to test a range of different engines?

Unfortunately, cost and timescales prevented a range of different engines being tested. Although only an XPAG engine was used, it proved to be an excellent choice, with the results providing a better understanding of the problems. As a result of this work, it is possible to suggest solutions that are relevant to the majority of petrol engines that use carburettors or early fuel-injection systems.

What is wrong with a rolling road?

Other people have asked: "Why go to the expense of fitting an engine into a test cell and not use a rolling road?"

The dynamometer and engine test cell at MACE provided the ultimate test environment. It was possible to fit a large range of instrumentation, including air-to-fuel ratio (AFR) meters, thermostats, vacuum gauges and exhaust gas monitors, etc. These enabled detailed measurement of engine performance. It was easy to change the fuel under test. Everything was accessible, allowing the engine to be tuned for every test run.

Figure 1-2 shows the engine behind an array of eight yellow temperature meters. These are connected to thermocouples on the fuel pump and carburettors. The AFR meter is on the left-hand side. At the top right are the temperature readings for the cooling water and exhaust gas.

1-2: Engine and measuring equipment.

1-3: Dynamometer.

1-4: Control room.

Figure 1-3 shows the water-braked dynamometer. This measured the engine's torque output. Its large gauge showed the engine revolutions per minute (rpm). The dynamometer was managed from the control room (Figure 1-4). This is where the gas analyser and various other readouts were located.

The tests

The main aims of the tests were to:
- Investigate why the engine was not running properly when using modern petrol.
- Look at the differences between brands and grades of petrol.
- Determine the effect of ethanol blended petrol.
- Measure the optimum ignition and carburettor settings for different grades of petrol and compare these with the original manufacturer settings.

Chapters 4 and 12 describe why, during the ignition cycle, the sparkplug needs to be fired before the piston reaches the top of its stroke. This is called 'ignition advance.' As engine rpm increases, it is necessary to also increase the ignition advance. The level of advance needed at each engine rpm is called the 'advance curve.' Running an engine either too advanced or too retarded not only reduces its power output, but can also cause serious damage. For each of the tests, the engine timing was manually adjusted for maximum power output, the optimum setting.

Chapters 4 and 5 describe why the air-to-fuel ratio (AFR) entering the engine must be precisely controlled by the carburettors. The XPAG is fitted with twin SU carburettors. These are variable jet carburettors, described in Chapter 5. They are virtually identical to the Stromberg carburettors fitted to other models of cars. One advantage of the SU is that it is very easy to adjust the air-to-petrol mixture. Again, for each test the carburettors were adjusted to give an optimum AFR.

The engine was tuned to give the best performance for each of the test runs. This ensured it was running optimally on each of the fuels, allowing direct comparison of the ignition timing and carburettor settings for each test. The recommendations made in Chapters 11 and 12 are based on the differences between these measurements and the original manufacturer settings.

For each test:
- The throttle position was set to either full ('Wide Open Throttle' or WOT), half or quarter open.
- The desired engine rpm was set using the dynamometer.
- The mixture was adjusted to give an AFR of 0.95. This corresponds to maximum power.
- The ignition timing was adjusted to give maximum torque output.

- Finally, the rpm was reset using the dynamometer.
- After a few minutes, when the engine had settled, the instrument readings were recorded.

This sequence was repeated over a range of rpm for each throttle setting and for each different fuel. This allowed a large amount of data to be collected for many different scenarios.

Measurements

The dynamometer gave a direct measure of the engine's torque output. A formula converted this to power. The exhaust gas analysers measured how well the fuel burned as described in Chapter 4. A timing light with an advance feature measured the actual ignition advance.

Chapter 5 describes how the carburettor suction piston height measures the volume of air flowing into the engine. A pointer and scale was fitted to the top of the carburettor to allow this to be measured. This is shown in Figure 1-5.

In order to identify the cause of the Hot Restart Problem, thermocouples were fitted to each of the two carburettors, the inlet manifold and petrol feed. These measured running temperatures and how they changed after the engine had been stopped. The white wires can be seen on figure 1-5, these connect the thermocouples to the yellow meters.

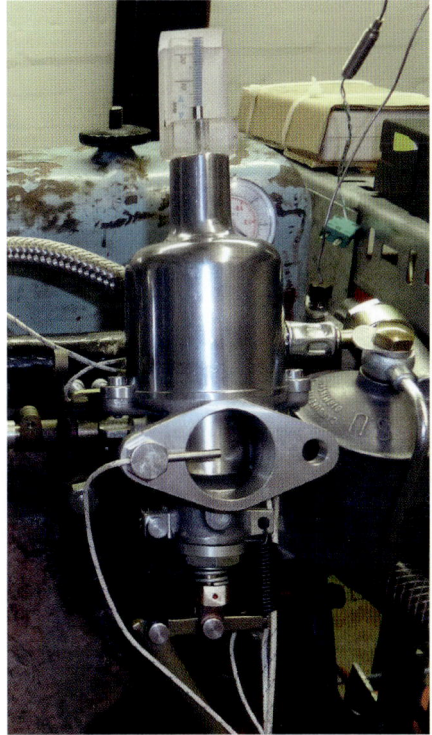

1-5: Instrumentation on the carburettor.

Chapters 4 and 13 describe why it is necessary to further advance the ignition timing on low throttle settings. Vacuum gauges were fitted to measure inlet manifold pressure (visible behind the top of the carburettor on the right-hand side) to allow the vacuum advance curve to be determined.

Test fuels

Unfortunately, it was not possible to get classic petrol as a comparison. Avgas was used in its place. This is used in light aircraft. It still contains tetraethyl lead as an anti-knock agent, and its composition is close to that of classic petrol.

The physical properties of petrol from the 1930s and 1960s were obtained for comparison and are presented in Chapter 2.

Three specialist fuels were provided by Anglo American Oil. These were tested in addition to a range of 95RON (85MON) and Super grade 98RON (88MON) petrol from local filling stations. Some of the road tests I had done, suggested adding kerosene (paraffin) to the petrol made my TC run better. A 20 per cent kerosene/petrol mix was included in the tests, with surprising results. Petrol with 10 per cent ethanol (E10) from France completed the list.

Other measurements

In addition to tests in the engine cell at Manchester, some petrol samples were sent for professional volatility analysis. These revealed the main reasons behind the Hot Restart Problem. These results are discussed in Chapter 2.

Finally, the E10 petrol was used in a set of over-winter tests to measure its corrosive effects on fuel system components. Results are reported in Chapter 3.

Summary

These tests are the most comprehensive ever performed on a classic engine using modern fuel. They delivered a large amount of comparative data, providing an in-depth understanding of the problems and potential dangers.

Chapters 7 and 8 describe these problems in more detail and Chapters 9 to 13 suggest solutions that can be employed to mitigate their effects.

2 Petrol volatility

Petrol is not a single substance. It is a mixture of over 300 different hydrocarbons and additives. This mixture in modern petrol is very different from that of classic petrol. More than just the removal of the tetraethyl lead. The different mixtures change petrol's physical properties and the way it behaves in an engine.

One of the biggest differences with modern petrol is how well it evaporates, otherwise known as its volatility. A greater volume evaporates at low temperatures compared to classic petrol. This is the cause of the Hot Restart Problem and the reason modern petrol 'goes off' when stored.

This chapter presents the volatility data for modern petrol and compares it to classic petrol. It discusses how the heat from the engine boils the petrol in the carburettors and why this causes the Hot Restart Problem, a problem that affects all classic vehicles with carburettors or early fuel-injection systems to one extent or another.

This higher volatility causes a second problem. If petrol is left in the tank of a car for any time, it will 'go off' as the more volatile fractions evaporate. This alters the petrol's characteristics and the way the engine runs.

Petrol volatility profile

Every hydrocarbon that makes up petrol has its own boiling point. Heat a sample of petrol and as the temperature rises the different hydrocarbons will boil away, the most volatile first. Once all this has evaporated, the temperature will rise until the boiling point of the next component is reached, and so on. Measuring the volume of fuel that evaporates as the temperature increases gives a 'volatility profile,' technically called a 'distillation curve.' The components evaporating at the lowest temperatures are called front end components. Those at the higher temperature, back end components.

The way an engine starts, ticks over, and runs depends on the shape of the distillation curve. Even though two fuels may have the same distillation curves does not mean they are the same.

Figure 2-1 shows the volatility profile for 95RON (85MON) petrol (blue line). This is compared to the curves for 1930s and 1960s classic petrol (the red and purple dotted lines). While there is little difference between the petrol sold in 1930 and that sold in 1960, they both differ from modern petrol.

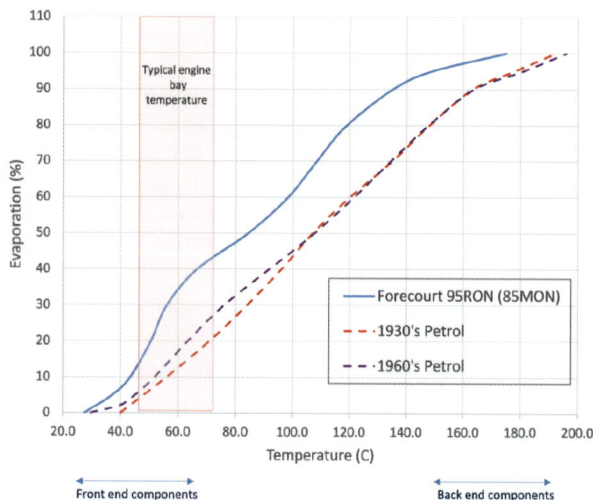

2-1: Evaporation v/s temperature.

At a temperature of 75°C (167°F) 45 per cent of modern petrol, nearly half, will have evaporated, twice the volume of classic petrol. This temperature is typical of those that may be found in the engine bay of a car.

Storage problems

If a car, classic or modern, is not used for a few weeks, the petrol stored in its tank can 'go off.' Owners have reported that their cars are difficult to start. When they finally do, they run rough until filled with a fresh tank of petrol. The front end components of modern petrol evaporate at ambient temperatures, changing its characteristics. Data from BP Australia Ltd produced in 2005, demonstrates this effect:

Time	1 Week	2 Weeks	3 Weeks	4 Weeks	5 Weeks
Per cent volume of petrol lost	3%	5%	8%	10%	15%
Density (gm/cc)	0.75	0.76	0.765	0.78	0.79

With fewer front end components, the petrol does not evaporate as readily; it is harder to start the engine. When it does start, the increased density will cause the engine to run rich and rough.

This creates a dilemma. One recommendation is to store a vehicle with a full tank of petrol to stop condensation. This is particularly important if the petrol contains ethanol. The dangers of water contaminating ethanol blended petrol are described in Chapter 3.

However, if you follow this recommendation, you will lose a greater volume of petrol through evaporation, and there will be more 'bad' petrol in your tank when you come to use your vehicle.

Suggestions to address the storage problem are:

· Only keep your tank half full when you store your vehicle.
· Add some fresh petrol before you try to start or use it after storage.
· Use a fuel that contains anti-oxidants, metal deactivators and corrosion inhibitors (possibly super grade petrol).
· Use a specialist storage fuel such as Sunoco Optima 98 (see Chapter 10).

Hot Restart Problem

Chapter 5 describes how a carburettor delivers an accurate air-to-fuel ratio (AFR) to the engine. Carburettors can only work with liquid petrol, not with petrol vapour. With over 40 per cent of modern petrol evaporating at typical engine bay temperatures, it is surprising a carburetted engine manages to run at all. Cars with older mechanical fuel-injected systems are also susceptible to vapour locks in the petrol.

Modern, fuel-injected cars do not suffer from this problem for two reasons. Firstly, the petrol in the engine bay is held under pressure, increasing its boiling point. Secondly, it is continually circulated between the engine bay and petrol tank, keeping it cool.

Engines and heat

A petrol engine produces colossal quantities of heat. Unfortunately, only around a third of this heat energy is converted into power to move the car forward. The remaining two thirds is wasted. A normally tuned XPAG engine produces 40 kilowatts of power at full throttle. This is one third of the heat energy from the petrol. The remaining two thirds waste heat, or 80 kilowatts, needs to be dispersed.

Approximately 55 per cent of this waste heat is lost via the exhaust, heating the exhaust manifold, exhaust pipe and vented as hot gasses. The cooling system removes 35 per cent of the heat. Eight per cent heats the engine block and oil, and the final two to five per cent is lost through other means. All this heat makes the engine bay very hot. Imagine a room in your house with 80 electric fires all switched on: it would certainly get very warm!

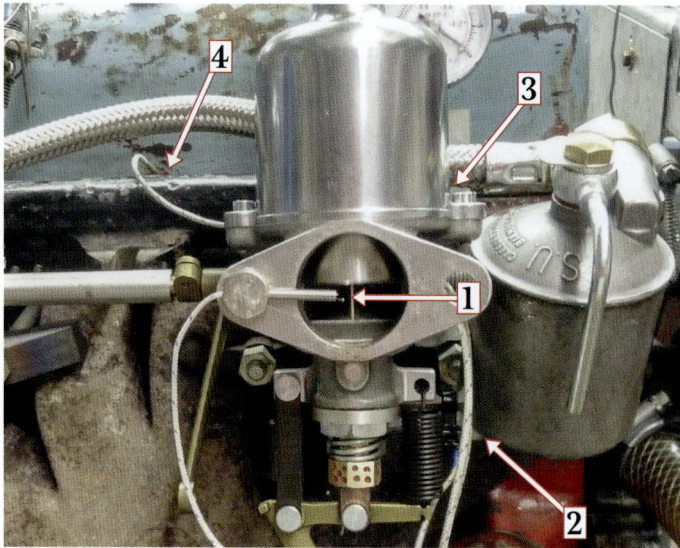

2-2: Thermocouples on the carburettors.

Temperature measurements

One objective of the tests at Manchester was to identify the causes of the Hot Restart Problem. To measure the temperature of the petrol, thermocouples were attached to the engine, carburettors and fuel pump. Figure 2-2 shows the position of three of the thermocouples fitted to each carburettor:

- The carburettor air inlet (1 on Figure 2-2). Typically this was 30°C (86°F), Representative of an engine on a warm day.
- In the transfer pipe connecting the float chambers to the carburettor body (2 on Figure 2-2 and Figure 2-3). Typically this was 42°C (107°F). This was surprisingly low considering this was positioned under 2.5cm (1in) away from the 400°C (700°F) exhaust manifold.
- Embedded into an aluminium gasket fitted between the carburettor and inlet manifold. These measured the heat conducted into the carburettors from the engine. Typically this was 36°C (97°F). These plates (3 on Figure 2-2 and Figure 2-4) were also fitted with vacuum gauges to measure inlet manifold pressures.

Two additional thermocouples were fitted to:

- The cylinder head between cylinders 2 and 3. Typically this was 170°C (340°F) heated by the exhaust gases passing through the cylinder head (4 on Figure 2-2).

- The fuel pump outlet to measure the temperature of the petrol flowing into the carburettors (not shown). Typically this was 22°C (72°F) room temperature, corresponding to a warm summer's day.

With the engine running at full power, the highest petrol temperature of 42°C (107°F) was in the transfer tubes. Even at this temperature, less than 10 per cent of modern petrol will evaporate – insufficient to cause any running problems. The low temperatures of the carburettors are shown more dramatically on Figure 2-5. This shows a thermal image alongside a picture of the carburettors with the engine running at 3000rpm at full throttle.

On the thermal image, the colour blue indicate low temperatures, yellow, red and white high temperatures. The blue 20°C (68°F) to 42°C (108°F) carburettors and float chambers are silhouetted against the white, hotter than 300°C (570°F), exhaust manifold.

Despite being so close to the exhaust manifold, the carburettors do not get hot. Nor does the cross

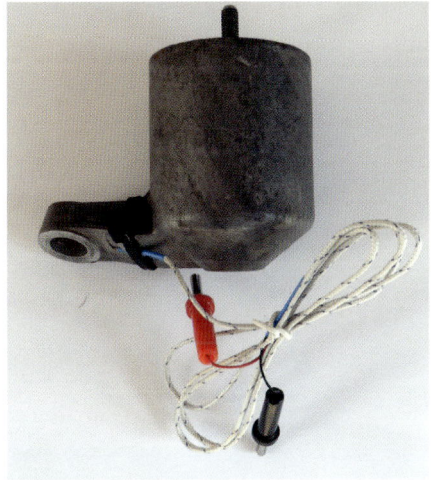

2-3: Float chamber and transfer pipe.

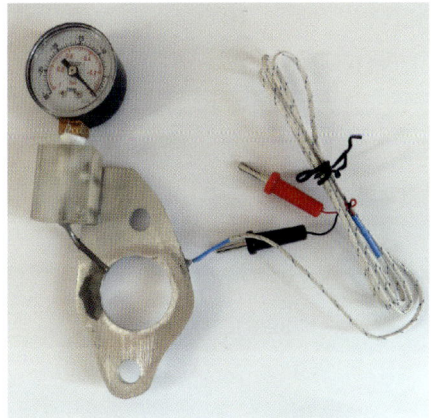

2-4: Plate between inlet manifold and carburettors.

fuel pipe that links the two carburettors and loops over the top of the hot exhaust manifold. This is surprising as there was no heat shield fitted between the exhaust manifold and carburettors. This arrangement represents the 'worse-case' scenario.

When the engine is running, especially under power, a large volume of petrol is flowing through the carburettors keeping them cool. This is why they are able to work with modern petrol.

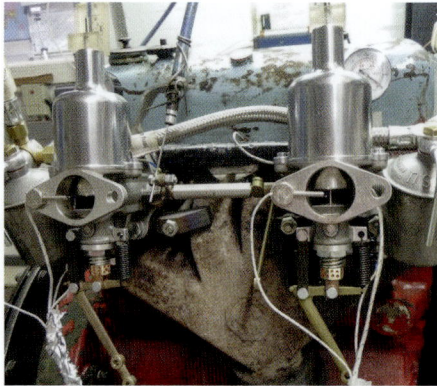

Suction chambers

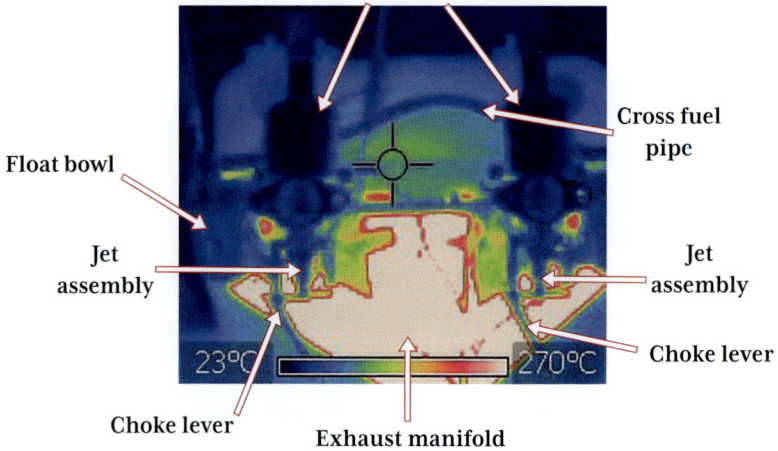

Cross fuel pipe

Float bowl

Jet assembly

Jet assembly

Choke lever

23°C   270°C

Choke lever   Exhaust manifold

2-5: Thermal image of the carburettors.

Overheating and the Hot Restart Problem

In slow moving traffic or when stopped, two effects work to increase the temperature of the petrol. Although the engine is running at low power and producing less heat, there is less air flowing through the engine bay. The rate of heat loss is reduced and engine bay temperatures will rise. Secondly, petrol is flowing slowly through the carburettors. It has more time to heat up. When the engine is switched off, heat 'soaks' out of the engine, exhaust and radiator. With no air flowing around the engine, temperatures will continue to rise.

The distillation curve for 95RON (85MON) petrol (Figure 2-1) shows a rapid rise in the volume of fuel evaporating between 45°C (113°F) and 75°C (167°F). This is typical of the temperature rise seen in the petrol when a vehicle is slow moving or stopped. As the petrol boils, the pressure of the vapour forces it out of the carburettor jet. Petrol collects in the inlet manifold temporarily flooding the engine. The vapour bubbles in the jet then result in a much weaker mixture. This is what causes the engine to stop or prevents it from starting.

The most reliable solution is to lift the bonnet (hood) and wait for the temperature to drop. Alternatively, it may be possible to nurse the engine back to life using the choke to enrich the mixture. Although it will run very unevenly, driving a short distance will bring cooler petrol into the carburettors from the tank. Increased airflow will help reduce engine bay temperatures. Both these will lower the temperatures to the point where less of the petrol evaporates and the engine returns to normal running. At this point the choke should be taken off.

Summary

The increased volatility of modern petrol at relatively low temperatures is the prime cause of the Hot Restart Problem. As temperatures around the engine rise in slow moving traffic or after the vehicle stops, petrol starts to boil. Carburettors cannot deliver the correct mixture if there are bubbles of vapour in the petrol and this is why an engine will stop.

There are four ways to address the Hot Restart Problem:
· Change the petrol you use.
· Decrease the amount of heat generated by the engine.
· Increase the heat removal from the engine bay.
· Reduce the amount of heat reaching the fuel system.
These are discussed in Chapters 9 to 13 and summarised in Chapter 15.

3 Ethanol blended petrol

In the UK, 95 octane and some premium brand fuels can have up to 5 per cent ethanol or alcohol blended into the petrol. This is referred to as E5. Petrol with 10 per cent ethanol (E10) or higher concentrations are sold in Europe and other parts of the world. These are usually marked on the pump as E10, E15, etc. Unfortunately, in the UK, you do not know if the petrol you are putting into your tank contains ethanol or not. If it does, you do not know at what concentration.

Adding ethanol to petrol is not new. Cleveland Discol was introduced in 1928 and sold until 1968. The manufacturers claimed it "contributed to a brilliant performance and better mileage, because it keeps the engines cooler and cleaner;" "the perfect cold-weather fuel." However, it is not known how much ethanol Cleveland Discol contained, making it difficult to compare with modern fuels. The good news is, after 40 years of use in what are now today's classic cars, Discol did not appear to cause failures of their fuel systems.

This chapter outlines the issues associated with the use of ethanol blended petrol. It also discusses the findings of the Manchester, and other, tests. No matter what the age of your vehicle, there is a paradox with the use of ethanol blended petrol. It gives benefits yet can be the cause of serious problems.

Why add ethanol to petrol?

Government policy to reduce carbon emissions from vehicles is driving the addition of ethanol to petrol. Adding ethanol reduces both carbon emissions and pollution. The carbon in the ethanol comes from renewable sources; it is a by-product of the sugar industry. The Manchester tests also showed it burns better than non-blended petrol, producing less pollution.

In the UK, the ethanol added to the petrol is subject to the normal fuel duty. There is a penalty called the Road Transport Fuel Obligation (RTFO) for oil companies that do not add ethanol. The producers have to pay a penalty of 30 pence per litre when ethanol levels are less than a set quota. This quota is set to rise in 2020. As a result, ethanol levels in UK fuel will rise.

One benefit of ethanol is that it boosts the octane rating, reducing the need for other additives. Typically methyl tertiary-butyl ether (MTBE) and ferrocene are added to petrol to improve the octane rating. These, in turn, replaced tetraethyl lead (TEL) after it was banned. Some brands of super grade fuel sold in the UK use ethanol to boost their octane rating.

If you are using an ethanol blended fuel, it is important NOT to try to remove the ethanol, as this will reduce its octane rating.

Why is octane rating important?

During the engine's compression stroke, the air/fuel vapour mixture in the cylinder is heated. If there are any incandescent carbon deposits, they can cause the mixture to autoignite. These create multiple ignition points before the sparkplug has fired. This causes a phenomenon called pinking or knocking. The engine makes a sound like a pebble being shaken about in an empty tin. Pinking can cause serious damage to an engine.

A fuel's octane rating is a measure of how resistant the petrol is to autoignition. The higher the octane number the less likely it is to autoignite and cause pinking. Contrary to some views, octane rating is not an indication of the burn speed of the mixture.

There are two similar measures for octane number. Research Octane Number (RON) and Motor Octane Number (MON). RON is determined by running the fuel in a test engine with a variable compression ratio. This is the measure widely used in Europe. MON is determined at 900rpm engine speed instead of the 600rpm used for RON. Usually, MON numbers are around six to ten points lower than RON. 85 MON is approximately the same as 95 RON. (MON is commonly used in America and Canada.)

How can you check if a petrol contains ethanol?

It is relatively easy to check if a fuel contains ethanol, but you must be VERY CAREFUL. Petrol vapour is highly flammable. Ensure there are no naked flames, sources of sparks or similar anywhere near when you test the fuel.

The test is based on one of the problems with ethanol blended petrol. When water is present, the ethanol will be sucked out of the petrol and into the water. This increases the volume of the water.

To perform the test, add some food colourant to a quantity of water. Pour it into a small bottle, such as a drinking water bottle and place it on a flat surface. Mark the level of the top of the water. Fill the bottle to the top with the petrol you want to test. Put the lid on and shake. After the coloured water has settled at the bottom, look at its level. If it is above the line, the volume of the water has increased, proving that the petrol did contain ethanol.

The two bottles in Figure 3-1 were filled to the same level with water that had been dyed brown. The difference in height of the water is easy to see. The petrol on the right contains ethanol, the one on the left does not. (Note: there is nothing at the bottom of the bottle, this is just an effect of the light on the indentation.)

When you have finished the test, remember to dispose of the petrol and water responsibly. Do not pour it down the drain!

3-1: Sample test bottles. Petrol without ethanol (left) and with (right).

What are the problems with ethanol?

There are four potential problems with ethanol blended petrol. It:

- Rots non-metallic components such as rubber hoses, seals, diaphragms and plastic floats.
- Contains oxygen which weakens the mixture.
- Is corrosive to metallic components such as the steel petrol tank and aluminium float chambers.
- Absorbs water and can create a separate acidic layer underneath the petrol.

The effects and seriousness of these problems are discussed in the following sections.

Rots rubber hoses, seals, diaphragms, etc

The ethanol in the petrol dissolves the plasticisers used in rubber and plastic components, making them brittle. Old braided fuel hoses that have lasted for many years have failed within weeks of being subjected to ethanol blended fuel.

Owners with vehicles that have electric fuel pumps are fortunate. These make it easy to detect most leaks in the fuel system. When you come to start the car from cold, switch on the ignition. Before you start the engine, listen to the petrol pump. An electric pump will normally click for 15 seconds or so, as it replaces fuel lost through vaporisation. This clicking should slow and stop. If it doesn't, you may have a leak in the fuel system or a problem with the petrol pump, leaking needle valves in the float chambers or a faulty float. In any case, this should be investigated.

Note: if you are trying to restart a hot engine, then a continued clicking may be due to petrol vapour in the fuel pump or hoses.

Is this tendency to rot hoses, seals, etc, a real problem? Probably not. These components are relatively cheap and easy to replace with ethanol tolerant parts. In older vehicles it is probably wise to replace them in any case due to their age.

It is important that owners are aware of this problem and keep a regular watch for any petrol leaks, especially if the fuel hoses are hidden inside a metal braiding. If your car is fitted with an electric fuel pump, make a habit of listening to the clicking. If not, as part of a regular service routine, check the petrol system components for leaks. BEWARE, should you smell petrol in an enclosed space such as a garage, vent the space as soon as possible. Make sure there is no possibility of flames or sources of electrical sparks before the petrol vapour has cleared. Petrol vapour is HIGHLY flammable.

Enleanment

Pure petrol is a mixture of hydrocarbons – it contains only carbon and hydrogen atoms. Hence, one litre (1000ml) of petrol contains 1000ml of hydrogen and carbon atoms. Ethanol contains 35 per cent oxygen atoms by weight. One litre of E10 (with 10 per cent ethanol) will only contain 965ml of hydrogen and carbon atoms.

In effect, you are being "short changed" when you fill up with ethanol blended petrol. All else being equal, a car that returns 30mpg will only give 28.9mpg when running on E10. At Manchester, the XPAG engine ran more efficiently on ethanol blended fuels. Hence, it is possible E10 will give a better mpg despite containing less hydrocarbons.

The replacement of hydrocarbons by oxygen has a second effect called enleanment. Carburettors deliver a precisely measured volume of fuel to a given volume of air (Chapter 5). The ideal is called the stoichiometric ratio with 14.7 parts of air to one part of petrol. At this ratio, when the petrol burns there are exactly the correct number of oxygen atoms to combine with the hydrogen and carbon atoms to produce water and carbon dioxide. Adding ethanol to the petrol has a double effect. Not only does ethanol reduce the number of hydrocarbons, it also increases the number of oxygen atoms in the mixture by a similar amount. This has the effect of weakening the mixture.

Carburettors are set rich as standard, and the tests at Manchester showed they will cope with E5 without any adjustment. If E10 is used, they should be adjusted to make the mixture slightly richer. With the older SU carburettors this corresponds to screwing the adjusting nut DOWN by one to two flats on the nut. On the HIF type the adjusting screw should be turned ¼ to ½ of a turn clockwise.

At higher concentrations of ethanol, variable jet carburettors will need slightly richer metering needles. Weber and similar fixed jet carburettors and mechanical fuel-injection systems will need larger jets and possibly different emulsion tubes. Modern computerised fuel-injection systems should adjust the mixture automatically.

Enleanment does not appear to cause any practical problems with ten per cent or less ethanol added to the petrol.

Corrosive to metallic components

Corrosion of metals is caused by two different processes:
- Oxidation – where oxygen combines with the molecules on the surface of the metal to create a metal oxide. This is most common when an acid comes into contact with a metal.
- Galvanic corrosion – occurs when two dissimilar, electrically connected metals are immersed in a liquid which is able to conduct electricity. This acts like a battery. As current flows between the metals, the positive anode corrodes. Galvanic corrosion is often the cause of electrical problems in classic cars when the connectors in the wiring loom build up an insulating layer of metal oxide.

Ethanol blended petrol causes both oxidation and galvanic corrosion.

Additives, recommended by the Federation of British Historic Vehicle Clubs (FBHVC), are available. These reduce the severity of oxidation. They may not be necessary as some brands of petrol claim they already contain such additives.

Fuel systems are made from a range of different metals such as brass, copper, steel and aluminium. These are often connected to each other and can conduct electricity. As a result, you get galvanic corrosion when using ethanol blended petrol.

Figure 3-2 shows the results of an over-winter test. The left-hand photograph is of a piece of aluminium stored for three months in ethanol-free petrol. This was connected to a similar sized piece of mild steel. The photograph on the right shows a piece of aluminium stored in the same way but in petrol containing 10 per cent ethanol.

The black dots on the right-hand photograph are galvanic corrosion pits. The horizontal and vertical scratch marks were caused by abrading the samples before the test to ensure they were free of contaminants.

Voltage measurements suggest the combination of stainless steel and aluminium will cause a higher level of galvanic corrosion in aluminium. Other metal combinations, such as brass/aluminium or brass/steel, appear to be far less affected.

While ethanol blended petrol causes galvanic corrosion, its effects are relatively minor compared to its other corrosive properties.

Galvanic corrosion to aluminium with steel

Ethanol Free **10% Ethanol**

3-2: Over-winter galvanic corrosion – ethanol free and E10.

Water absorption

Ethanol blended fuel absorbs water vapour from the atmosphere. When a certain concentration of water is reached, the ethanol/water mix will separate from the petrol. It will form a layer underneath the petrol, as shown in the test bottles in Figure 3-1. In practice, it is unlikely you will experience this problem. The greatest risk arises if your vehicle has been stored in a damp environment for a long time.

What is a more serious threat is the ingress of water in liquid form. Fuel systems of classic vehicles are far from waterproof. The filler cap is often located where rain can easily get in. Sometimes the float chambers on the carburettors have ticklers on them that can also allow water in.

It only requires one droplet of rain to enter a tank of ethanol blended petrol to cause a far greater corrosion problem. The droplet will not mix with the petrol. It will fall to the bottom of the tank where it will absorb ethanol. Water containing ethanol is highly corrosive. If the car is not being used, this droplet of water will sit in the same place, corroding the base of the fuel tank or float chamber.

Figure 3-3 shows a water-filled test container which has been used to store pieces of steel and aluminium for six months. The container was filled with water that had been pre-mixed with ethanol blended petrol. You cannot see the pieces of metal. They are hidden by the rusty water.

3-3: Over-winter storage container – ethanol and water.

3.4: Before …

… and after.

Figure 3-4 shows the severity of the corrosion. The top photograph shows the steel and aluminium before the test. The bottom photograph is the two samples after they were removed from the container. The degree of corrosion of both the steel and aluminium is extreme.

This problem occurs in practice. Figure 3-5 shows similar corrosion inside one of my float chambers, possibly caused by rain entering through the tickler pin when driving in heavy rain.

The dark line about 12 o'clock on the bottom of the chamber is caused by the corrosion. When the float chamber was opened, there was what looked like a worm, sitting underneath the petrol. This was almost certainly a small quantity of water that had absorbed ethanol.

Figure 3-6 shows the petrol

3-5: Corrosion in float chamber.

removed from the float chamber. The rusty brown droplet of water can be seen at the bottom with some debris from the petrol tank.

Once the water settles at the bottom of the petrol tank or float chambers, it will continually absorb ethanol from the petrol. Becoming more acidic and corrosive over time. This is a far greater threat than galvanic corrosion. Furthermore, this problem will occur regardless of the concentration of ethanol.

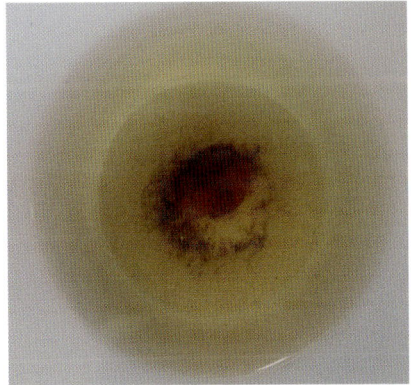

3-6: Petrol removed from float chambers.

The degree of galvanic corrosion increases with ethanol concentration. In contrast, corrosion from water which has absorbed ethanol will be just as corrosive with E5, or petrol with lower ethanol concentrations.

Unfortunately, the additives sold to protect fuel systems against ethanol give no benefit. While they will mix with the petrol, they will not mix with the ethanol/water mixture.

The problems caused by water getting into fuel systems is not new. Ethanol blended petrol makes the water significantly more corrosive and these problems more severe.

It is the greatest threat of using ethanol blended petrol.

Hot Restart Problem

Ethanol has a low boiling point of 78.4°C (173°F). Some people have suggested it may make the Hot Restart Problem worse. The tests at Manchester showed this is not the case. By 80°C (176°F), 48 per cent of a sample petrol without ethanol had evaporated, compared to only 40 per cent of a super grade petrol with ethanol. The petrol with the ethanol was less volatile. However, the reduction in volatility may be due to other factors, for example, it being a super grade blend.

The brand and grade of petrol is most likely to affect the severity of the Hot Restart Problem rather than its ethanol content.

The tests showed that ethanol blended petrol burned more efficiently than non-blended petrol. This is discussed in Chapter 10. As a result, for a given load, the exhaust gas temperatures will be lower. This will create less heat in the engine bay, reducing temperatures and possibly the effects of petrol vaporisation.

Solutions

The only practical solution to avoid the problems caused by ethanol blended petrol is to use a fuel known to be ethanol free, such as Sunoco Optima 98 (Chapter 10).

If you are using pump fuels in the UK, you do not know if they are ethanol blended or not. The composition of a particular brand or grade may change day on day. For example, two samples of 95RON (85MON) petrol bought in Manchester at the same filling station within days of each other were different. One sample contained ethanol, the other did not.

Take great care to avoid getting water (eg rain water) into either the petrol tank or carburettor float chambers. Perhaps it is worth draining the petrol tank and float chambers once a year and allowing them to dry out to ensure no water/ethanol mix remains.

Possibly the most satisfactory solution is to copy modern practice. For the tests at Manchester, I was sent drums of petrol and replacement parts for modern carburettors. In both cases they were coated to prevent corrosive damage. Coatings protect against both oxidation and galvanic corrosion.

Slosh coat the inside of the petrol tank with an ethanol proof paint. Unfortunately, I am not aware of any similar products that can be used to coat the inside of the float chambers.

Summary

While there are issues, it appears that ethanol blended petrol is not the 'baddie' that some people fear.

There are two practical problems. Rotting petrol hoses and seals and the severe corrosive effects of any water that may get into the petrol. On the positive side, the tests at Manchester showed the engine ran better on ethanol blended petrol (Chapter 10). Rotting hoses is something that can occur in older vehicles. Age as well as ethanol causes degradation of fuel hoses, etc.

The second, water in the petrol. This can be avoided with care. Slosh coating fuel tanks will significantly reduce this risk.

Stainless steel fuel tanks may not be the answer. They can be attacked by acid such as the water/ethanol mix. They also may cause more severe galvanic corrosion to any aluminium parts. For example, the tank petrol level sender. It may be worth getting replacement stainless tanks slosh coated.

Ethanol blended petrol is here to stay. Over time, concentrations of ethanol will rise. This chapter should reduce an owner's worries and help them be better prepared for the future.

4 Suck, squeeze, bang & blow

Search on the internet for information on how petrol engines work and you will find the answer: "there are four cycles, induction, compression, ignition and exhaust." Or as described in this chapter, suck, squeeze, bang and blow. However, the operation of these engines is more complex than this simple description. To help the reader appreciate the results of the Manchester tests, this chapter introduces some of the concepts affecting the combustion of fuel in a four-stroke spark-ignition engine. It describes the journey taken by a single cylinder in an engine running at 3000rpm. While the valve timings apply to an XPAG, they are virtually identical to any petrol engine, old or new.

Our piston completes the four stages of the cycle in 40 thousandth of a second (40ms). Think how fast one second is, and imagine that 1ms was the same as one second. On that timescale, one minute would last 17 hours! It would take around 11 hours for our cycle to complete. 1ms is so fast that even gases act like solids.

Suck

The start of our journey is when the piston is at top dead centre (TDC). At this point you might expect the exhaust valve to snap shut followed by the inlet valve opening as quickly. Valves cannot open and close instantaneously. Delays in opening the inlet valve reduces an engine's power. The engineers who designed these engines knew valves could start to open earlier or close later than expected.

At the start of our journey, the inlet valve will already have begun to open. It started to open 0.6ms (11° before TDC). The exhaust valve is still open. It will take another 1.3ms (24° after TDC) before it closes.

The 1.9ms when both valves are open is called valve overlap. This is beneficial at higher rpm.

At the top of the 'blow' stroke, the piston has expelled most of the exhaust gases. As the inlet valve starts to open, a 'scavenge' effect takes place. The rush of gases into the exhaust port draws in air/petrol mixture through the inlet valve.

At TDC the cylinder is not empty. The 45.5cc combustion chamber (about 15 per cent of the 312.5cc cylinder volume) still contains 1200°C exhaust gases from the previous cycle. As the piston starts the 'suck' stroke, these will continue to vent through the exhaust valve until it closes 24° after TDC. The remaining hot gases will cool as they expand. If you ever studied Physics, you may remember Boyle's Law: as a gas expands, it cools, and when compressed it gets hotter.

As the piston falls it will reach the point where the pressure in the cylinder becomes lower than the inlet manifold pressure. The air/fuel mixture will start to flow into the cylinder. Induction has begun.

The volume of mixture entering the cylinder is controlled by the throttle butterfly. This is a brass disc that pivots when the throttle is pressed. As it rotates, it reduces the area of the restriction in the inlet manifold allowing more mixture to flow into the engine. As more air flows through the carburettor, the suction piston responds. It moves upwards withdrawing the tapered needle from the jet. This allows more fuel into the air stream. The way the SU or variable jet carburettor works is described in more detail in Chapter 5.

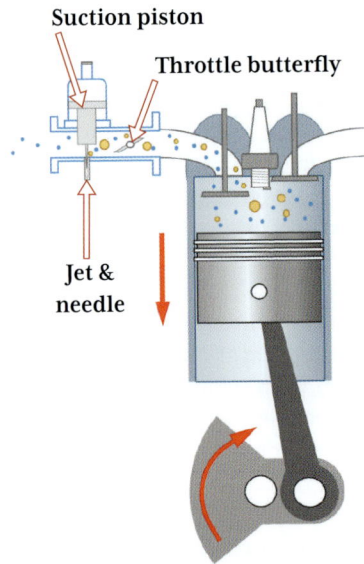

Suction piston

Throttle butterfly

Jet & needle

4.1: Suck.

To get the greatest power from an engine, the 'suck' cycle needs to induct as much air/fuel mixture as possible. In a normally-aspirated engine, the volume of mixture entering the cylinder depends on engine capacity. As petrol vapour occupies about 14 times the volume of liquid petrol, the more liquid petrol that can be inducted, the greater the power output.

Superchargers or turbochargers increase the pressure in the inlet manifold, forcing more mixture into the cylinder. This is why these engines generate more power than normally-aspirated engines with the same capacity.

Depending on throttle setting and engine rpm, around ten per cent of the petrol will evaporate in the carburettor, cooling it. The remaining 90 per cent will enter the engine as different sized droplets of liquid petrol.

In normal road use, when the engine is running at part throttle, the volume of petrol evaporating in the carburettor will not be noticeable. However, for those who want to maximise power output, it reduces the overall volume of petrol entering the engine and hence its power. Additionally, the cooling effect can also cause the carburettors to ice up, especially on cold, damp days. This is more likely in engines with exposed carburettors such as motorbikes.

As discussed in Chapter 2, modern petrol has more front end components than classic petrol. These evaporate at lower temperatures. This has two negative effects on carburetted engines. Firstly, it increases the volume of petrol evaporating in the carburettor. Less liquid petrol is inducted, reducing power output. Secondly, it increases the cooling of the carburettors, increasing the risk of icing.

The first air/fuel mixture entering the cylinder meets the residual hot exhaust gases. These heat the incoming mixture evaporating some of the petrol droplets and cooling the residual gasses in the

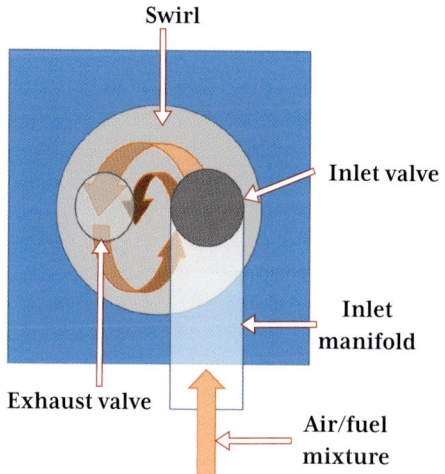

4-2: Cylinder swirl.

process. Even though these are extremely hot, they will not contain enough energy to evaporate all the inducted petrol.

The inlet valve on many two-valve per cylinder engines is offset to the side of the cylinder. The advantage of this is that it causes the inducted mixture to swirl during the suck stroke. This both helps the petrol droplets disperse in the air and increases turbulence, something that is very important during the 'bang' cycle.

Before petrol can burn, it must be a vapour. To get the optimal mixture for the bang cycle, all the liquid entering the cylinder must evaporate or boil and mix with the air. This boiling is unlike that in a kitchen kettle used for water. In a kettle, bubbles form in the bulk of the liquid. Because the suck cycle is so fast, the petrol in the cylinder can only evaporate molecule by molecule from the surface of the droplets. The small droplets with a large surface area relative to their volume, evaporate the fastest.

Squeeze

After reaching bottom dead centre (BDC), the piston starts to rise. Induction continues for another 3.2ms (57° after BDC) until the inlet valve closes. During this time the air/fuel mixture entering the cylinder some 90mm above the piston, does not feel the effect of its upward motion. The mixture continues to flow into the cylinder. This increases the cylinder pressure about 0.2 to

0.5lb/in^2 (140 to 350kg/m^2) above that of the inlet manifold. This is called the stagnation pressure.

The 'squeeze' stroke does not start in earnest until the inlet valve has closed. It continues for another 6.8ms until the piston reaches TDC, after 20ms or half way into our journey.

During this stroke, the pressure and temperature of the mixture increases. This provides extra heat to evaporate more liquid petrol.

Pockets of rich and weak mixture will form as the liquid petrol evaporates. Turbulence in the gases disperses these pockets as the squeeze stroke progresses.

In common with many modern designs, the combustion chamber in the cylinder head of the XPAG is boat shaped. As a result, the outside

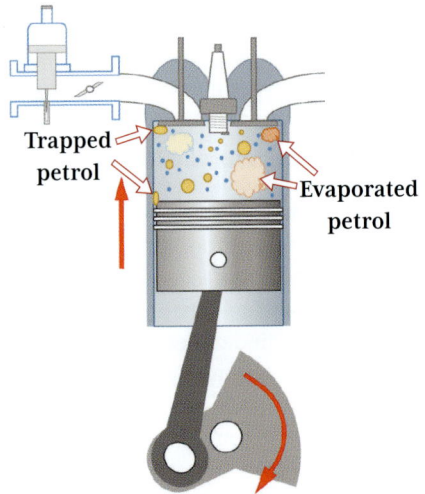

4-3: Squeeze.

edges of the cylinder head overlap the bore. As the piston approaches the top of the stroke, the gases on the outside edge are 'squished' into the combustion chamber, increasing turbulence and mixing.

At the end of the compression stroke, liquid petrol may still be present in the cylinder. This can be trapped between the piston and cylinder wall, around the valves, or as large droplets of fuel that have not evaporated. Even if the carburettor is delivering the correct mixture, the presence of liquid petrol results in a weak mixture at the end of the stroke.

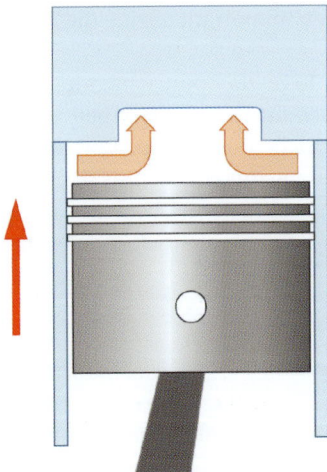

4-4: Squish.

Bang

There are two ways the vapour of a flammable liquid can ignite. The flash point is the lowest temperature at which an ignition source, such as the sparkplug, can ignite the mixture. The lowest temperature at which it will spontaneously ignite, burning without a source of ignition, is called the autoignition temperature.

Autoignition is very bad for an engine. It causes the pressure in the cylinder to rapidly increase, resulting in pinking or knocking. This is a mechanical tinkling sound that occurs typically at full throttle and low rpm. It sounds like pebbles being shaken about in an empty tin. An ideal fuel has a low flash point and high autoignition temperature. The higher the octane rating of the fuel, the higher the autoignition temperature.

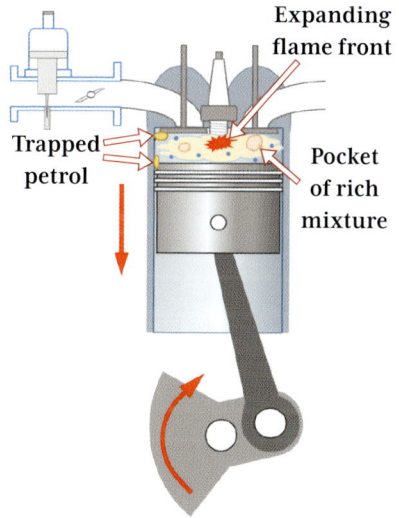

4-5: Bang.

The XPAG, in common with many older engines, has a low compression ratio of 7.25:1 compared with 9:1 or higher in modern engines.

In the 1950s, improved petrol quality allowed compression ratios to be increased. With higher compression ratios there is more heating of the air/fuel mixture during the squeeze stroke, and more liquid petrol will evaporate. Its final volume is smaller, allowing it to burn more efficiently. These engines produce more power than lower compression engines of the same capacity. The main disadvantage is that the engine is more prone to pinking or knocking caused by the mixture autoigniting during the compression stroke.

Combustion is initiated by an electrical spark in the gap of the sparkplug.

After the sparkplug fires, combustion takes place in three phases:

- A fireball of burning mixture, about the diameter of a human hair, is created between the electrodes of the sparkplug. On the timescale of the running engine, this flame front expands exceptionally slowly.
- Once the fireball has grown to about the size of the sparkplug electrode, turbulence takes over. This spreads the ignition points, causing the remaining mixture to burn very rapidly.

- As the temperature in the cylinder rises, any remaining liquid petrol is vaporised. These temperatures are sufficiently high to burn any hydrocarbons in the cylinder. Including engine oil that may have leaked past the valves or piston rings.

As petrol burns, the hydrocarbon chains break down; the hydrogen (H) combines with the oxygen (O_2) in the air to produce water (H_2O). The carbon (C) combines with the O_2 to produce carbon dioxide (CO_2). Each litre of petrol liberates a huge 33 million joules of heat energy – enough to boil 100 kettles of water. The heat energy from the burning fuel increases the temperature of the gases to over 2000°C. The theory Boyle's Law states that as the temperature of a fixed volume of gas increases so does its pressure. As the pressure increases, it forces the piston down the cylinder. This is how the engine converts the heat energy in the petrol into power.

As the piston falls so does the temperature and pressure of the gases. Greatest power is produced when the pressure in the cylinder reaches its peak at 17° or 0.9ms after TDC.

After the sparkplug has fired, it takes approximately 2.6ms for the mixture to burn and reach peak pressure. To provide enough time for this to happen, the sparkplug needs to be fired before the piston reaches the top of its stroke. This is called ignition advance. Simplistically, the time to reach peak pressure is constant and independent of rpm. Hence, as rpm increases, and the piston is moving faster, the sparkplug needs to be fired earlier in the cycle to give the same time for the fuel to burn. A graph showing the ignition advance against rpm is called the advance curve.

Typically, at 3000rpm, the sparkplug is fired 30° before the piston has reached TDC. A race begins. As the flame front grows and the pressure in the cylinder rises, it is working against the piston, which is still moving upwards. At 30° advanced, the piston still has 8.5mm to travel (9.3 per cent of the stroke) before it reaches TDC. Knocking or pinking causes rapid increases in cylinder pressure before the piston reaches TDC. This puts an excessive load on the piston and big end bearing causing damage.

On 1930s and earlier cars ignition advance was set manually, typically by a lever on the steering wheel. On later cars this is done automatically by bob weights in the distributor. These fly out as engine rpm increases.

The growth of the initial fireball also depends on the pressure of the air/fuel mixture in the cylinder. This is mainly dependent on throttle setting, not, as may be expected, compression ratio. At light throttle settings, cylinder pressure is low and the growth of the flame front slower. To give more time for the air/fuel to burn, it is necessary to further advance the ignition timing. On later cars this is done by a vacuum pod on the distributor. This is connected to the inlet manifold to measure its pressure. This, in turn, is a measure of throttle setting. Earlier cars do not have a vacuum advance.

Maintaining the correct ignition advance is important. Running an engine too advanced will result in pinking or knocking and damage to the piston and big ends. Running too retarded increases the exhaust temperature, resulting in burned pistons and exhaust valves and damage to the cylinder head.

The 'bang' stroke suffers from a problem called 'cyclic variability.' The time taken for the air/petrol mixture to burn critically depends on a number of factors. These include the air-to-fuel ratio (AFR) around the sparkplug, the level of turbulence in the gases, etc. Even with a precisely timed spark, minor differences in these factors cause the timing of the peak pressure to vary on each cycle. This is discussed in Chapter 6.

Unfortunately, the power stroke ends all too quickly, 7.1ms after TDC when the exhaust valve starts to open (52° before BDC). The high pressure gases rush out of the cylinder in a process called 'blowdown.' Blowdown utilises the remaining combustion pressure to "get the gas in the exhaust moving." Without this effect, energy would be lost during the exhaust stroke as the piston would have to push the gases out of the cylinder.

Each piston is powering the car forward for just 18 per cent of the time of each cycle.

Blow

If the combustion were perfect, a mixture of water vapour, carbon dioxide and nitrogen rushes into the exhaust, expanding and cooling to around 500°C. The piston reaches BDC, 2.9ms after the exhaust valve started to open. As it rises, the remaining exhaust gases are pushed out of the cylinder.

Unfortunately, combustion is not perfect; petrol vapour requires oxygen to burn. Droplets of petrol that vaporise late in the cycle leave pockets of poorly mixed vapour. Although the temperature is sufficiently high to cause these to

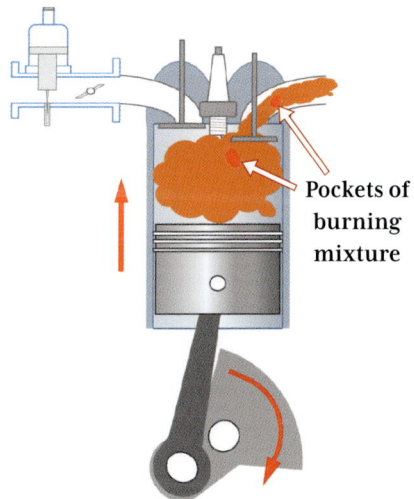

Pockets of burning mixture

4-6: Blow.

burn, the absence of oxygen means they will not burn properly. They result in unburned hydrocarbons or carbon monoxide in the exhaust gases. As the unburned hydrocarbons travel down the exhaust system they may mix with oxygen. As they burn they further increase exhaust temperatures.

A gas analyser reveals a great deal about the combustion process. The unburned hydrocarbons show how much petrol has been unable to burn in the cylinder or exhaust due to a lack of oxygen. This can arise either because of a rich mixture or poor combustion as described above. NOX or nitrous oxide, (NO) is produced at high combustion temperatures when the nitrogen in the air oxidises. NOX is bad for three reasons: it uses energy, reducing the engine's efficiency; it reduces the amount of oxygen available for the fuel to burn; and it's an atmospheric pollutant. The presence of NOX is usually an indication of high ignition temperatures caused by a weak mixture.

In contrast, high levels of unburned hydrocarbons or carbon monoxide indicate insufficient oxygen. This is either due to a rich mixture or poor combustion.

What about modern cars? While the valve and ignition timing will differ slightly from an XPAG, the journey described above is very similar. There are three main differences:

- The fuel is injected as very small and evenly sized droplets, typically 50μm diameter (about the size of a human hair), a fifth of the size of those produced by a carburettor. These not only mix with the air more effectively, caused in part by the careful design of the inlet manifold, they evaporate faster. Ultimately, this creates a more evenly distributed mixture of air and petrol vapour in the cylinder before the ignition fires.
- Compression ratios are higher than in classic cars, increasing the compressive heating. There is more energy to vaporise the petrol.
- The ignition timing is continuously adjusted to ensure that the mixture burns optimally. These engines are typically far less advanced than the XPAG. As a result, the race between the piston and flame front is shorter and no longer left to chance.

At 3000rpm, our journey ends after 40ms only to start over again. Each cylinder completes the cycle described above 25 times per second. As you can see from this brief description, it is far more complicated than just induction, compression, combustion and exhaust.

5 Carburettors

Chapter 4 described the four cycles of a spark-ignition engine, outlining the part the carburettor plays. This chapter describes the operation of the carburettor in more detail.

There are two types of carburettor: fixed jet, such as Weber and Solex; and variable jet, such as SU and Stromberg. Both these types work in similar ways.

There is a narrowing in the carburettor restricting the air flowing into the engine. This is called a choke. There is a fine jet in the choke that is connected to a reservoir of petrol held at atmospheric pressure. If you ever studied Physics, you may remember Bernoulli's principle. As gas flows through a restriction or choke, its speed increases and its pressure drops. The reduced pressure in the choke draws the petrol out of the jet into the air stream.

Their names suggest the difference between the fixed and variable jet carburettors.

Fixed jet carburettors have a fixed sized choke and jet. Opening the throttle allows more air to flow through the choke. This decreases the pressure in the choke, drawing more fuel into the air stream. Unfortunately, this arrangement delivers a weak mixture at low throttle settings and a rich one when the throttle is open wide. Fixed jet carburettors have extra jets to add petrol at low throttle settings. They also have an emulsion tube to weaken the mixture at high throttle settings.

In contrast, variable jet carburettors have a variable choke and jet. These are able to deliver an accurate air/fuel mixture over a range of throttle settings. There is no need for extra jets or emulsion tubes.

The XPAG engine is fitted with twin variable jet SU carburettors. These have a number of advantages over fixed jet carburettors. Firstly, it is very simple to adjust the volume of petrol delivered by the carburettor. This allowed them to be tuned to deliver the optimum air-to-fuel ratio (AFR) for each of the tests. Secondly, it is possible to measure the volume of air flowing through the carburettor while the engine is running. This provided the means of comparing the performance of different fuels during the tests.

This chapter describes the way the variable jet carburettor works. Its aim is to give owners confidence when maintaining their cars through a basic understanding of their operation. It also helps the reader in understanding some of the results of the Manchester tests.

This chapter uses the 1¼in HS2 SU carburettors fitted to the XPAG engine as an example. Even so, the principles apply to all types of variable jet carburettors such as SU and Stromberg.

SU carburettor

SU Carburettors are a marvel of engineering, first designed by George Herbert 'Bert' Skinner. It was first produced in 1908 by his younger brother Thomas Carlyle 'Carl' Skinner. Unchanged in the way they operate, the Skinners Union or SU carburettors were fitted to production cars until 1993.

5-1: Originally from the book Skinner's Union published by The SU Carburettor Company Ltd. Reproduced with the author's permission.

The original SU carburettor (Figure 5-1) had the connection to the inlet manifold at the top with a sloping suction chamber on the left. Modern SU (Figure 5-2) and Stromberg (Figure 5-3) carburettors have their connection to the engine on the side and a vertical suction chamber. Apart from this, the suction chamber and the float chamber are recognisable on this original SU carburettor.

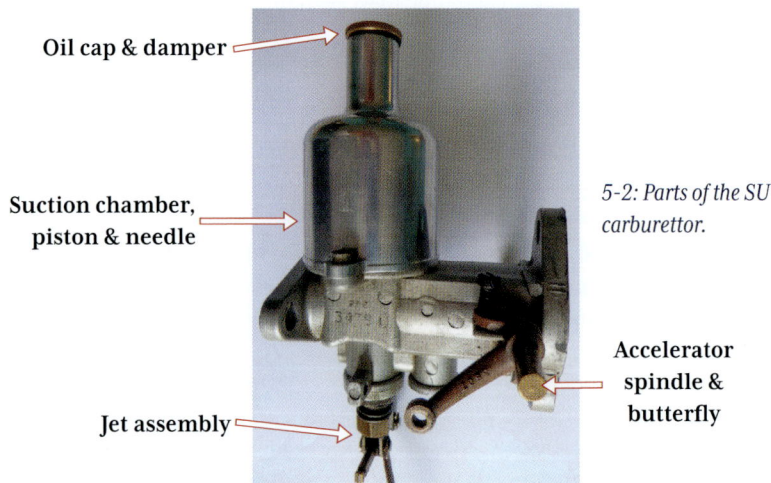

Oil cap & damper

Suction chamber, piston & needle

Jet assembly

5-2: Parts of the SU carburettor.

Accelerator spindle & butterfly

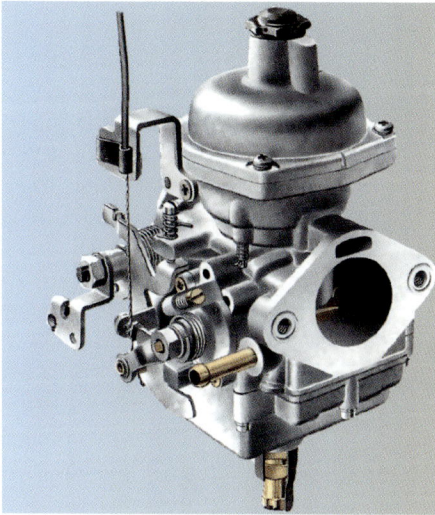

5-3: Stromberg carburettor.

Improved engineering in 1914 allowed the leather bellows, fitted to the early carburettors, to be replaced with a close-fitting suction piston. While SU patented the use of a close-fitting piston, they did not patent their earlier design using the leather bellows. With the advent of modern rubber, this enabled Zenith to produce the Stromberg carburettor in competition to SU.

The most obvious difference between the SU and Stromberg carburettors is the size of the suction chamber. In all other ways they are virtually identical.

Operation – what does a carburettor do?

In operation, the carburettor measures the volume of air flowing into the engine. It mixes this with a metered volume of atomised petrol droplets to give a precise air-to-fuel ratio (AFR). It needs to achieve this for all throttle settings and engine rpm, as well as over a range of atmospheric temperatures and air pressures.

With non-ethanol blended petrol the theoretical AFR is 14.7:1, ie for every 1g of petrol you need 14.7g of air. With this ratio all the hydrogen atoms will burn to produce water (H_2O), and all the carbon atoms will burn to produce carbon dioxide (CO_2). The maximum power is achieved with an AFR of between 12.5:1 and 13.5:1, having an excess of petrol or a richer mixture. This is because some of the carbon burns to carbon monoxide (CO). This uses only one oxygen atom rather than the two needed when it burns to carbon dioxide (CO_2). Hence the need for more carbon atoms and less oxygen.

Normally, the term lambda is used to describe the mixture. This is the ratio of the AFR the carburettor is producing divided by the theoretical ideal of 14.7:1. Numbers less than one correspond to a rich mixture and greater than one indicate a weak mixture.

When tuned, the SU maintains an accurate lambda between 0.85 (richer) and 0.95 (maximum power) over a very large range. From a low airflow at tick over, to full throttle at 5000rpm or more, when some three cubic meters of air are flowing through it every minute.

The SU carburettor is a volumetric device. It measures volumes of air and petrol. AFR is defined as a mass ratio. The volume of a given mass of petrol or air depends on its density. The density of petrol is defined by the producers and changes very little. However, the density of air can vary with ambient temperature, barometric air pressure or altitude. Fortunately, the SU is relatively insensitive to these changes, allowing our cars to work in the cold of winter, the heat of summer, at sea level and when driving over the top of alpine passes.

What is amazing is that it achieves all this with only one moving part!

For many, the SU carburettor is something to be left untouched. Once set up, it will continue to work for many miles, but it does benefit from regular maintenance. Changes in modern fuel means they may need to be re-tuned to deliver the best performance.

How does an SU carburettor work?

The two functions of the SU or Stromberg carburettor are to control and measure the volume of air flowing into the engine. The butterfly valve, connected to the accelerator, controls the volume. Measurement is achieved by a piston fitted inside the suction chamber. Sitting on top of the carburettor body, this is the most recognisable feature of these carburettors.

The carburettor inlet can be thought of as consisting of four different areas (Figure 5-4). These are labelled P1 to P4 under the diagram.

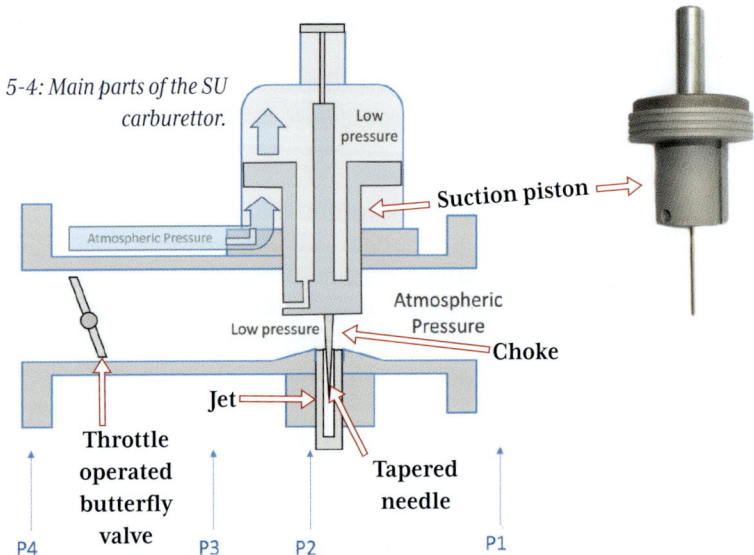

5-4: Main parts of the SU carburettor.

Low pressure

Suction piston

Atmospheric Pressure

Low pressure

Atmospheric Pressure

Choke

Jet

Throttle operated butterfly valve

Tapered needle

P4   P3   P2   P1

44

- P1 – where the air flows into the carburettor. This is at atmospheric pressure.
- P2 – the area below the suction piston. This is called the choke. Here, the air is at a lower pressure than atmospheric.
- P3 – the area beyond the piston and before the throttle butterfly.
- P4 – the area beyond the throttle butterfly. On small throttle settings the pressure in this area can be a near vacuum. On Wide Open Throttle, it is virtually at atmospheric pressure.

Measuring the volume of air

In area P2, the bottom of the suction piston partially chokes the airflow into the engine. The lower the piston, the greater the choking effect. The pressure in the choke is less than atmospheric pressure. In practice, this low pressure area extends up and around the rear of the piston into an area called the vena contracta. There is a drilling through the suction piston connecting the vena contracta to the top of the chamber. This means the space above the suction piston is at the same low pressure as the choke.

Without any seals that could cause friction, the top of the piston is a close fit inside the suction chamber. It is free to move up and down. In the case of a Stromberg carburettor, the piston is supported by a rubber diaphragm, again allowing it to move freely up and down. The bottom of the suction chamber is open to atmospheric pressure. When air is flowing through the carburettor, there is a pressure difference across the top and bottom of the suction piston. This creates a force causing the piston to move upwards.

As the piston moves up, the choking effect is reduced. A reduction in the choking effect, increases the pressure in the choke. In turn this reduces the pressure difference across the top and bottom of the suction piston. Hence, as the piston moves upwards, the force pushing it up drops. As a result, it can rise to a height where the force pushing it up is equal to its weight.

Should more airflow through the carburettor, the pressure in the choke drops. This increases the pressure difference across the piston, causing it to move further upwards until equilibrium is re-established. Conversely, as the airflow falls, so does the piston.

There is a constant pressure difference between the choke and atmospheric. This is independent of the volume of air flowing into the carburettor and equates to the weight of the piston. Hence the reason variable jet carburettors are referred to as constant pressure devices.

Strictly, this statement is not true. On early (pre 1950s) carburettors, the piston was made of bronze or brass. The ones in the 1¼in HS2 SU carburettor weigh 8.5oz (240gm). Later models had aluminium pistons with a steel insert to give the same weight. In these carburettors the downward force is fixed and independent of the height of the piston. They really are constant pressure devices.

In the 1950s, the heavier pistons were replaced with lighter aluminium pistons and a spring. The spring increases the downwards force to around 8.5oz (240gm). Stromberg carburettors also have a spring to increase the effective weight of the piston. As the piston rises, this compresses the spring, increasing the effective weight of the piston. The implications of this are discussed in Chapter 11.

Metering the volume of petrol

Connected to the bottom of the piston is a tapered needle that fits into a fixed diameter jet. The petrol level in the jet is controlled by the float chamber. This is at atmospheric pressure. As the pressure in the choke is lower than atmospheric, the petrol is forced out of the jet into the choke. The volume of petrol leaving the jet is controlled by the pressure difference and the size of the annulus between the tapered needle and jet. The petrol breaks into droplets as it enters the high velocity air stream in the choke.

The height of the piston is a direct measure of the volume of air flowing through the carburettor. The profile of the needle's taper is such that the correct volume of petrol enters the choke for every piston height. In this way, the variable jet carburettor delivers the correct AFR as finely atomised droplets into the inflowing air.

During the tests at Manchester, the height of the suction pistons in the carburettors was measured. This allowed the volume of air flowing into the engine to be compared for the different fuels. It was one of the key measurements in identifying some of the problems caused by modern fuel.

Damper and float chamber

The two other auxiliary parts that make up the SU carburettor are the damper and float chamber.

The guide rod in the centre of the suction piston is hollow. From 1947, SU carburettors were fitted with a damper. This is connected to the screw cap at the top of the suction chamber. It serves two purposes.

Firstly, it slows the rate the suction piston can rise when the throttle is quickly depressed, enriching the mixture to improve acceleration. Secondly, it stops the piston vibrating due to rapid pressure changes in the manifold when the inlet valve opens.

It is important that the oil in the hollow piston guide rod is checked and topped up if necessary. The oil should be filled to a level just below the top of the hollow rod. When the damper is replaced, a slight excess will overflow the top of the hollow rod and lubricate the guide. Overfilling the damper will cause the excess oil to enter the suction chamber, causing the piston to stick.

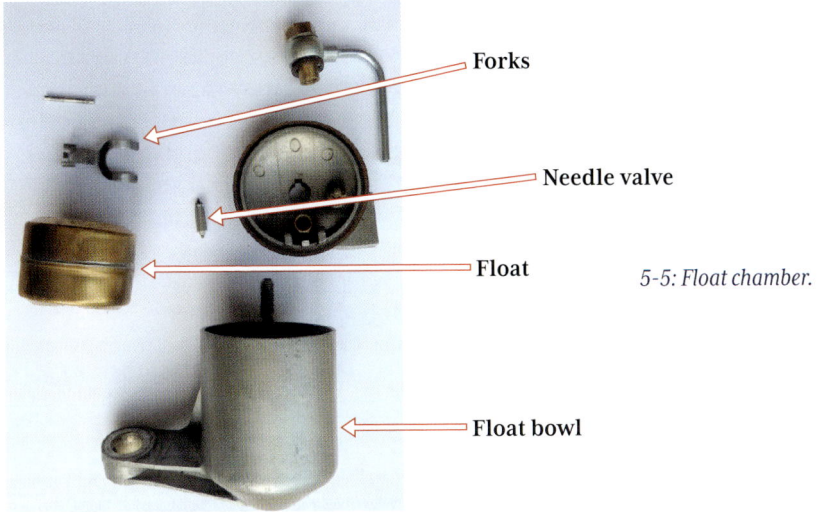

Forks

Needle valve

Float

Float bowl

5-5: Float chamber.

The sole purpose of the float chamber (Figure 5-5) is to keep the petrol level in the jet constant. As the fuel level drops, the float also drops, allowing the needle valve to open and more petrol to flow into the float chamber. The float chamber is connected to the jet and they act like a U-shaped tube with the petrol in each arm at maintaining same level. When no air is flowing through the carburettor, the level of petrol in the jet is the same as that in the float chamber.

Adjusting variable jet carburettors

There are a number of ways the operation of a variable jet carburettor can be adjusted. Chapter 11 describes how to tune the carburettor for modern fuel. The appendix describes how to rebuild and retune twin SU carburettors.

Damper oil

The viscosity of the oil in the damper has an impact on acceleration. The more viscous the oil, the richer the mixture during hard acceleration and the better the pickup. If the oil is too thick, it will stop the carburettor responding to slower throttle changes. If your car is fitted with a damper, it may be worth experimenting with different viscosity oils. Choose the one that gives the best acceleration response and smoothest running at low rpm. Normally, 20w/50 engine oil is recommended.

Jet height

Adjusting the height of the jet has the effect of changing the mixture over the whole rpm/throttle range. Moving the jet down increases the size of the annulus around the jet and tapered needle. This makes the mixture richer by allowing more fuel into the air stream. Moving it upwards reduces the size of the annulus, making the mixture weaker. (NOTE: on some Stromberg carburettors, the jet is fixed and the same adjustment is made by moving the needle up or down).

The jet (or needle height) is either set with a nut on earlier SU carburettors or a screw on later models and some Stromberg carburettors. Tolerances on the needle and jet are very tight. When adjusting the jet or needle height, only make changes by turning the nut one flat or screw ⅙ of a turn at a time.

On early SU carburettors, when the choke control on the instrument panel is pulled out, it drops the jet. This makes the mixture richer to allow cold starting.

Replacing the tapered needle

The other adjustment to the mixture is achieved by replacing the tapered needle. While this is a simple operation, with hundreds of needles having different tapers, retuning the carburettor is not a simple task. This should be left to the experts. Fitting a needle with a different taper alters the mixture at each rpm and throttle setting.

Burlen Fuel Systems Ltd sell books for both SU and Stromberg carburettors listing the needles and their profiles. They give the needle diameter in ⅛in steps down from the shoulder. Points near the shoulder affect the tickover and low rpm, low throttle settings. Points further down, higher rpm, higher throttle settings. A lower diameter gives a richer mixture.

Petrol height in the jet

The petrol height in the jet is set by adjusting the height of the float in the float chamber. If the petrol is too high, it can overflow the jet and make the mixture far too rich. Too low results in poor atomisation and dispersion of the petrol when it leaves the jet. This is discussed in more detail in Chapter 11.

Spring in the suction chamber

Later SU and Stromberg carburettors are fitted with a long spring in the suction chamber to increase the effective weight of the piston. It is possible to change these springs. In a similar way to changing the needle this alters the mixture at each rpm and throttle setting. The effect of this is discussed in more detail for SU carburettors in Chapter 11.

Maintenance

It is advisable to regularly maintain carburettors to ensure they are working effectively.

For both SU and Stromberg carburettors:
- Apply a little light oil to the throttle spindles.
- Clean out any debris or water from the bottom of the float chamber. This is very important if you are using ethanol blended fuel (see Chapter 3).
- Check the suction piston is free to move by unscrewing the damper. With extreme care, this can be used to lift the suction piston. Let it fall. It should drop back with a 'clunk.' With two or more carburettors, the pistons should all fall at the same rate. This is particularly important with SU carburettors where oil or dirt in the suction chamber can cause the piston to stick.
- Check the oil in the dampers.

If a piston in an SU carburettor is sticking it is advisable to clean it. For cars with twin carburettors, service one carburettor at a time and do not interchange the parts:
- Remove the suction chamber and clean the inside with white spirit. There is no need to oil.
- Clean the suction piston. Be very careful not to bend the needle.
- Replace the suction chamber in the same orientation as it was originally. Check the piston rises and drops easily.

Summary

The SU and Stromberg carburettors are a marvel of engineering and, if properly maintained, will give many years of service. The tests at Manchester suggest there are minor tweaks that will improve the way the engine runs on modern petrol. These are described in Chapter 11.

6 Combustion and cyclic variability

There is a popular view that modern petrol burns more slowly and hotter than classic petrol. This is not true. It is a symptom of a more complex phenomenon called 'cyclic variability,' introduced in Chapter 4.

This chapter describes cyclic variability in more detail. It explains the peak pressure frequency histogram, and discusses the effect of cyclic variability on the way the engine runs and why modern petrol appears to burn hotter and slower.

Cyclic variability affects all spark-ignition engines, classic and modern. A high degree of variability is bad for an engine. It reduces power output and causes damage to the piston, crankshaft bearings and valves. The Manchester data showed that the test engine suffered from a high degree of variability, particularly below 3000rpm. It identified some of the factors that increase the magnitude of the cyclic variability. These are introduced in Chapter 8, and ways to reduce this effect are discussed in Chapters 10 and 11.

Ignition

During the bang (or combustion) stroke of a spark-ignition engine, the interactions between the petrol, petrol vapour, air, and growing flame front are very complex.

After the sparkplug fires, the mixture burns in three phases. A fireball of burning mixture, the diameter of a human hair, is created between the electrodes of the sparkplug when it fires. The flame front in an air/fuel mixture at room temperature and atmospheric pressure expands at approximately 35cm/sec (14in/sec). On the timescales of an engine running at 3000rpm this is very slow. At this speed, in the time it would take the flame front to expand to 1cm (0.4in) in diameter, the engine would have nearly completed one revolution!

After the compression stroke, the temperature of the air/fuel mixture is around 300°C. Its pressure is about three times atmospheric or greater, depending on the throttle setting. Under these conditions, the flame front speed increases to approximately 50cm/sec (20in/sec); faster but still very slow.

Fortunately, the combustion of the mixture is not dependent on the speed of the flame front. As the fireball grows, turbulence mixes the burning gases with the unburned mixture. This rapidly increases the combustion speed. Even so, the initial growth from the diameter of a human hair to 2.5mm (1⁄10in), is very slow. To put this into perspective, 2.5mm (1⁄10in) is only the diameter of the electrode of the sparkplug. It takes around 3.5ms or 32° of engine revolution at 3000rpm for the flame front to grow to this size.

As well as depending on temperature and pressure, the speed at which the fireball grows depends on the air-to-fuel ratio (AFR) of the mixture.

Chapter 4 described how the ideal inducted air/fuel mixture is a stoichiometric ratio consisting of 14.7 times the mass of air to petrol. A more commonly used figure is lambda, the AFR equivalence ratio. This is the actual AFR divided by the ideal stoichiometric value of 14.7:1. Lambda values less than one correspond to a rich mixture, greater than one a weak or lean mixture.

Figure 6-1 shows the flame front speed in petrol as a function of lambda.

The highest velocity occurs when the mixture is slightly weak. It becomes slower when the mixture is richer or weaker. At a lambda of 0.90-0.95, where the engine is delivering maximum power, the speed has already dropped by 10-20 per cent.

The way the flame front speed varies with lambda is shown in Figure 6-1, It is virtually the same for any air/hydrocarbon vapour mixture. This is the reason the actual burn rates of classic and modern petrol are the same.

If the mixture of the air and petrol vapour around the sparkplug is too rich or too weak, this slows the growth of the initial fireball. It takes longer for the charge to burn. The turbulence that spreads the flame front through the mixture is a random effect. Just like the weather, which changes from day to day, the turbulence of the gases in the cylinder change on each combustion stroke.

Both these effects alter the time taken for the petrol vapour to burn on each cycle of the cylinder.

Pressure variations

The different burn rates result in differing pressure profiles in the cylinder for each combustion cycle.

Figure 6-2 shows the cranked pressure (ie no combustion) and the pressure curves for five individual combustion

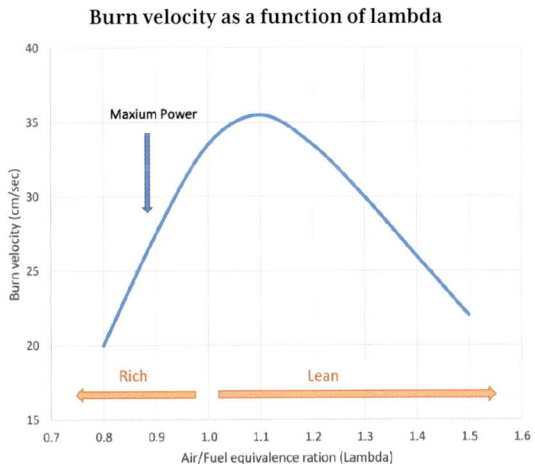

Burn velocity as a function of lambda

6-1: Flame front speed.

cycles in a test engine; both the crank angle at which peak pressure occurs and its maximum value vary.

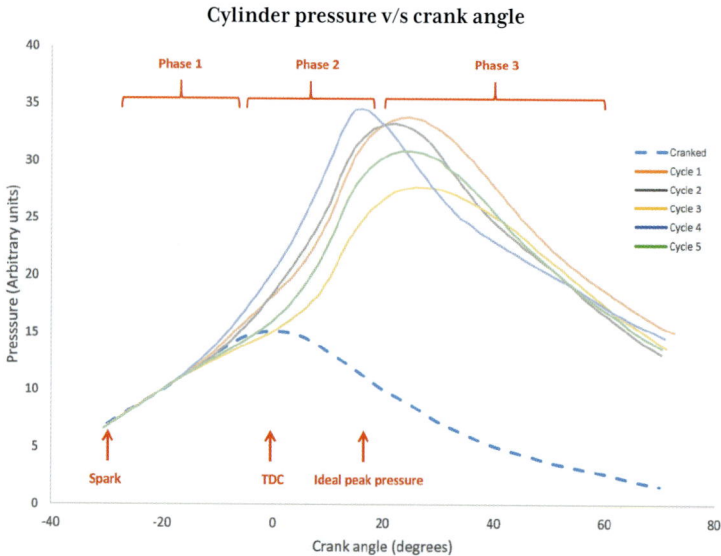

6-2: Cyclic variability.

For example, cycle 3 (yellow) peaks later and produces less power than cycle 4 (blue).

These differences occur because of variations in the combustion process. They are not caused by external factors such as wear on the distributor changing the ignition timing.

The differences between each cycle is called cyclic variability.

A peak pressure frequency histogram for these five cycles is shown in Figure 6-3. This is produced by counting how many cycles occurred between 15° and 19° after TDC, 20° to 24°, and 25° to 29°.

The peak pressure frequency histogram shows the number of cycles that occur in each angular range.

If it were possible to measure the timing of the peak pressure in a running engine over many cycles, the peak pressure frequency histogram would look like Figure 6-4. This curve is the result of a theoretical calculation.

Cycles whose peak pressure occurs in the red area on the left-hand side would cause pinking or knocking. Those in the red area on the right-hand side

Frequency plot of cycles

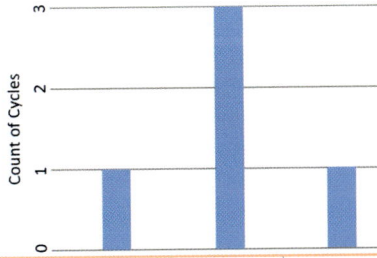

6-3: Frequency plot.

Count	1	3	1
Cycles in range	Cycle 4 ——	Cycle 1 —— Cycle 2 —— Cycle 5 ——	Cycle 3 ——
Angle of peak pressure after TDC	15° to 19°	20° to 24°	25° to 30°

Peak pressure frequency

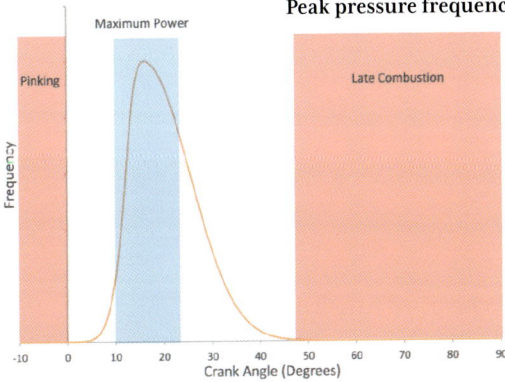

6-4: Low cyclic variability frequency plot.

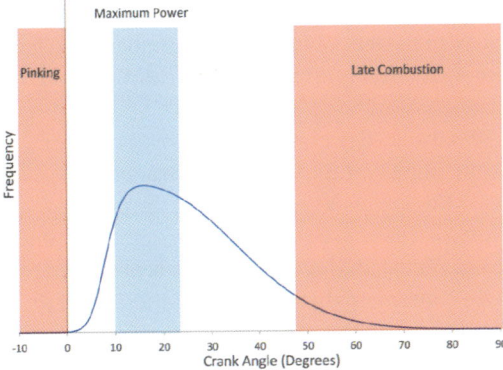

6-5: High cyclic variability frequency plot.

would have insufficient time for the mixture to fully combust before the exhaust valve opens. Only those cycles reaching their peak pressure in the blue area deliver the maximum power output.

This frequency plot represents the ideal case. Cyclic variability has the effect of spreading this curve out as shown in Figure 6-5.

Comparing Figures 6-4 and 6-5 two things are clear. Firstly, in Figure 6-5 more cycles occur in the red area on the right-hand side. A high degree of cyclic variability results in more cycles burning slowly. Secondly, fewer cycles occur in the blue area. The power output of the engine is lower; it runs less efficiently.

It is the higher number of late-combusting cycles compared to classic petrol that makes it appear as though modern petrol burns more slowly.

Evidence

Measuring the timing of the peak pressure in the cylinder of a running engine requires specialist equipment. This was not available at Manchester. The above curves are the result of a theoretical calculation to show the effects of cyclic variability. This raises the question, how do we know this is a real problem?

The way physicists prove a theory is by comparing its predictions with measurements. If the measurements match the theory, it is assumed the theory is correct. Some data from Manchester provides the means to assess the degree of cyclic variability in the XPAG engine.

The following section describes this test and shows how well the data fitted the theoretical predictions.

Measurement

The power output of an engine relates to the timing of the peak pressure. A cycle whose peak pressure is between 10° to 25° after TDC produces more power than one that occurs earlier or later. Retarding the ignition timing moves the peak pressure frequency curve, shown in Figure 6-5, to the right. Advancing the ignition timing moves the curve to the left. This alters the number of cycles that deliver the maximum power. This effect is illustrated in the three graphs in Figure 6-6.

When the ignition timing is 10° advanced (too retarded), a small number of cycles occur at the optimum time, shown by the blue area. Advancing the ignition to 20° (the ideal) increases the size of the blue area and the power output. At 30° advance (too advanced), fewer cycles occur at the optimum time but more than when the timing was set to 10°. Hence less power is produced than with the ignition set to 20° advanced but more than with the ignition set to 10°.

By measuring the power output of the engine at different ignition advance settings, it is possible to infer the number of cycles that deliver peak power. That is the size of the blue area shown in Figure 6-6.

Figure 6-7 shows the simulated power output based on low and high degrees of cyclic variability. These are compared with the measurements made at Manchester. The test used 95 octane, ethanol free petrol from a local filling station.

In Figure 6-7, the measured power output (orange points) are a good match to the theoretical simulation with a high level of cyclic variability (the blue line). When the engine is over advanced (the right of Figure 6-7), the power output is greater with high levels of cyclic variability than with a low level of variability. This is exactly what the data shows.

The comparison between the blue and grey curves show how a high degree of cyclic variability reduces the engine's power. In addition, maximum power occurs 2° to 3° later for the blue curve than the grey one. Chapter 12 suggests ways ignition timing can be adjusted to partially offset the effects of cyclic variability.

Similar measurements can be made on a rolling road, allowing the degree of cyclic variability to be estimated for any car. At a fixed rpm and throttle setting, record the engine's power output as the

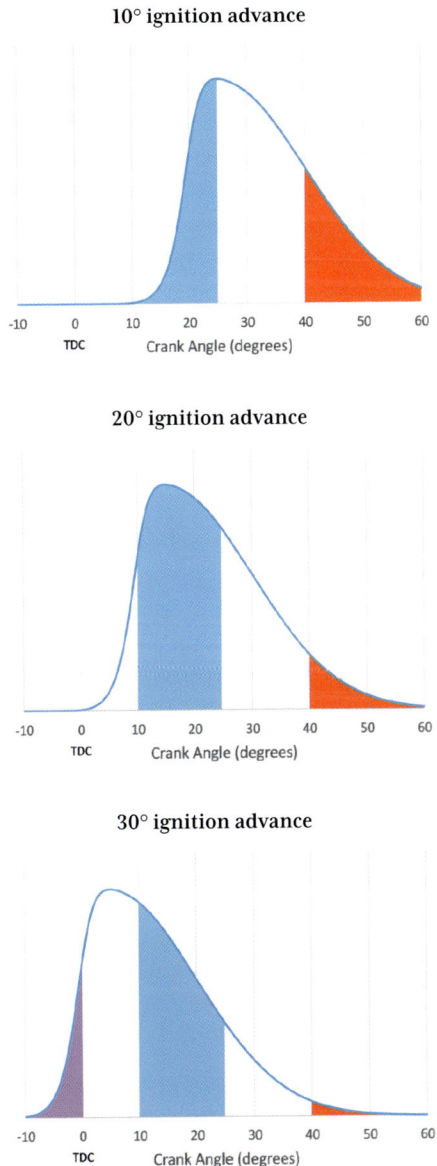

10° ignition advance

20° ignition advance

30° ignition advance

6-6: *Effect of advancing ignition timing on power.*

ignition timing is changed in steps from over-retarded to over-advanced. When plotted, this will give a curve similar to the one shown in Figure 6-7. The less the power is reduced as the ignition is advanced past the optimum timing, the worse the level of cyclic variability.

Torque as a function of ignition advance

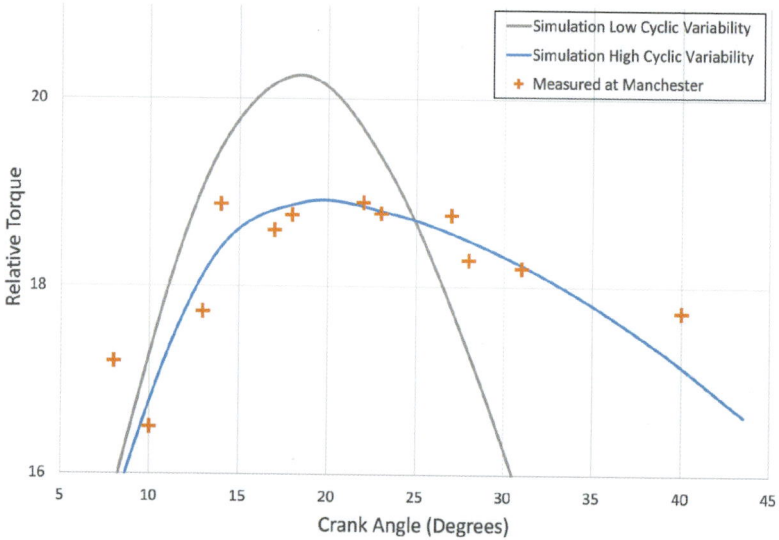

6-7: *Effect of advancing ignition timing on power.*

Why is cyclic variability so bad?

A high degree of cyclic variability is bad for an engine.

As Figure 6-7 shows, it reduces the power output. More throttle is needed to maintain a speed or accelerate at a given rate, the engine has to work harder. It operates less efficiently producing more heat. This increases the engine bay temperatures making the Hot Restart Problem worse.

It increases the number of cycles where the peak pressure occurs before TDC. For these cycles, the pressure on the piston rising up the bore is much greater. This can damage the piston and big end bearings. It can also trigger pinking or knocking, making the potential damage worse. Because the number of early cycles are only a small percentage of the total, it is impossible to hear the pinking or knocking.

The peak pressure for a large number of cycles occurs late increasing the exhaust gas temperature. The overly hot gases from these cycles can burn the piston and exhaust valves and crack the cylinder head. As Chapter 8 shows, this effect is at its worst when driving on public roads at low rpm, and with light engine loads.

Unfortunately, late combustion may also cause pinking or knocking. The hot gases can heat carbon deposits on the piston crown or cylinder head. These may be hot enough to cause the next cycle to pre-ignite. If your engine is pinking or knocking, it may not be due to the ignition timing being too advanced.

Figure 6-8 also demonstrates the effect of advancing the ignition timing on exhaust gas temperature. As the timing is advanced, the number of late-combusting cycles decreases, as will the exhaust temperature. This effect is demonstrated by the Manchester data. Figure 6-8 shows a marked drop in exhaust temperatures as the ignition timing is advanced.

Exhaust temperature as a function of ignition advance

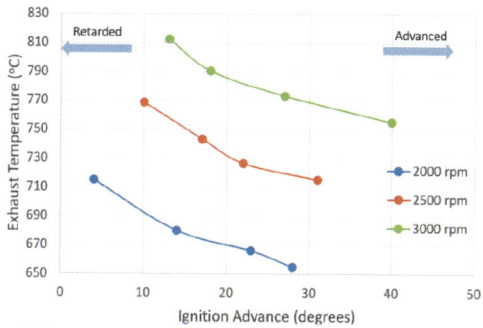

6-8: Effect of advancing ignition timing on exhaust temperature.

Summary
A high degree of cyclic variability is bad for an engine. The tests at Manchester demonstrated the XPAG experienced this effect, particularly when running below 3000rpm. Fortunately, by the choice of fuel and by tuning the carburettor and distributor, it is possible to mitigate this effect. These recommendations are described in Chapters 9 to 13.

7 Results from the tests – weak mixture

This chapter presents the findings from the SU carburettor data. These measurements were only possible because of the way it operates.

Chapter 5 describes the operation of the variable jet SU carburettor. It outlines how it measures the volume of air flowing into the engine and uses a tapered needle to deliver a precisely metered volume of petrol. One of the objectives of the tests at Manchester was to determine the optimum needle profile for modern fuel. Specifically to determine if it was different from the original manufacturer recommendations.

While the results from these tests are specific to the XPAG engine fitted with twin SU carburettors, as can be seen, they are applicable to most carburetted engines.

An understanding of the way SU carburettors work is not necessary, but appreciating one aspect of their operation is important. The height of the suction piston is a direct measure of the volume of air flowing through the carburettor. This piston is connected to a tapered needle which controls the volume of petrol. The air-to-fuel ratio (AFR) delivered at any given suction piston height is fixed, unless some other factor affects the carburettors' operation.

Standard needle profile

As more air flows through a variable jet carburettor, the suction piston rises, withdrawing a tapered needle from a jet. At any point, the diameter of the needle determines the volume of petrol entering the engine.

Needle profiles for SU and Stromberg carburettors are identified by two or three letters.

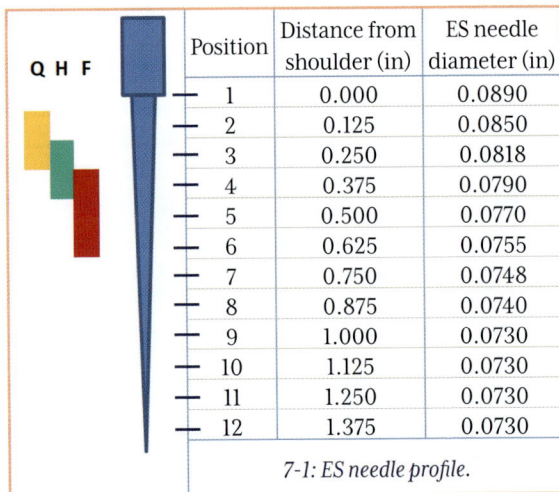

Position	Distance from shoulder (in)	ES needle diameter (in)
1	0.000	0.0890
2	0.125	0.0850
3	0.250	0.0818
4	0.375	0.0790
5	0.500	0.0770
6	0.625	0.0755
7	0.750	0.0748
8	0.875	0.0740
9	1.000	0.0730
10	1.125	0.0730
11	1.250	0.0730
12	1.375	0.0730

7-1: ES needle profile.

The profile of a needle is defined by its diameter at ⅛in steps down from the shoulder. The standard needle for the twin 1¼in SU carburettors fitted to the TB, TC, and TD is an ES needle. Its measurements are shown in Table 7-1. This also shows the approximate range of positions on the needle when running at quarter (Q), half (H) and full throttle (F) up to 3750rpm.

The greatest piston height is only 0.6in (position six) at 3750rpm, full throttle. Driving on the public highway at normal traffic speeds and throttle settings typically only uses the first four or five positions on the needle.

Measurements at Manchester

To determine the best needle profile, an indicator and a scale were fitted to the top of the suction chamber of each carburettor (Figure 7-2). This allowed the height of the piston to be measured while the engine was running. The faces of the jet-adjusting nut at the bottom of the carburettor were engraved with one to six dots, making it easy to determine how many flats it had been screwed down. The jet-adjusting nut can be seen at the bottom of Figure 7-2 showing one red dot.

Tests were run using different fuels, engine revs and throttle settings. For every test the carburettors were set to deliver a lambda of 0.95 using the jet-adjusting nuts. In effect, they were re-tuned for every test run.

The combination of the piston height and adjusting nut position allowed the ideal needle diameter to be calculated for each test.

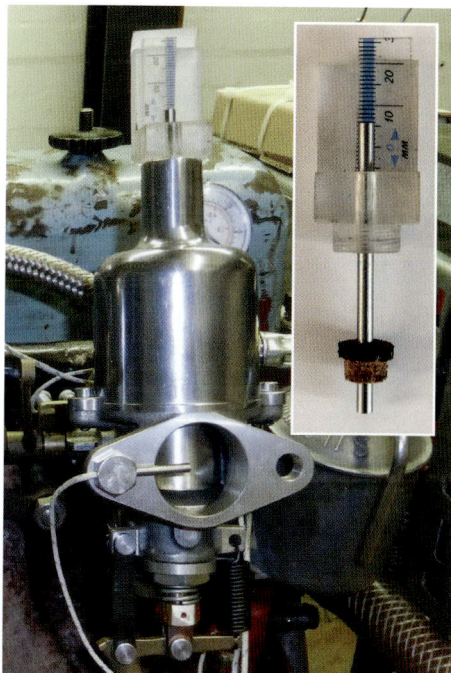

7-2: Piston height measurement.

Running lambda

Figure 7-3 shows a theoretical plot of how lambda would be expected to vary with carburettor piston height. Each dot represents one throttle setting/rpm combination. The white horizontal band is the ideal range for lambda. Lambda values greater than 0.98 represent a weak mixture, less than 0.83 a rich mixture.

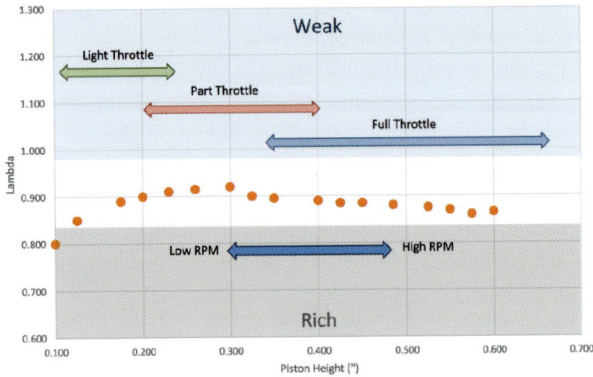

7-3: Expected piston heights.

The piston height is a measure of the volume of air flowing into the engine. Simplistically, at 2000rpm full throttle and 4000rpm half throttle, the same volume of air will enter the engine. The suction piston height would be the same for these two cases. As a result there is an overlap between throttle settings and piston heights. This overlap is shown in Figure 7-3.

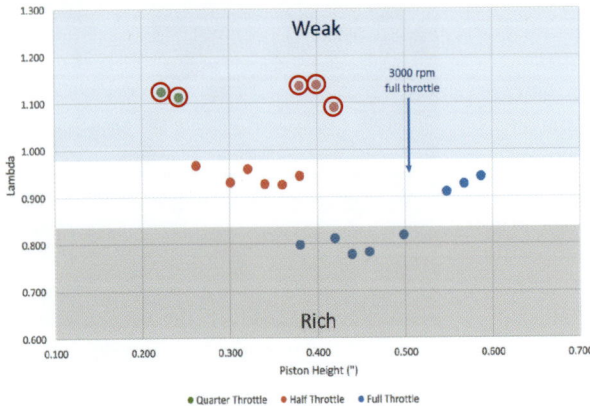

7-4: Measured piston heights.

The theoretical plot does not show a fixed lambda value. It presents an ideal profile; rich at light throttle to give a steady tick over (left-hand side of the graph). The ideal value at normal part throttle/mid rpm running, typical of driving on the road for economy, becoming richer as throttle setting and rpm increase (right-hand side of the graph). Under high loads, the unburned fuel helps protect the engine by cooling the exhaust valves.

Even though lambda was set to 0.95 for each test at Manchester, it is possible to calculate lambda had the engine not been re-tuned. Figure 7-4 shows this data for the engine running on ethanol free, 95 octane petrol, using the manufacturer recommended ES needle.

Weak mixture

Comparing Figures 7-3 and 7-4, they look very different. Figure 7-4 shows five points, circled in red, which stand out. For these tests, two quarter throttle (green) and three half throttle (red), the carburettors were delivering a very weak mixture. If it had not been re-tuned, the engine would not have started.

These tests were run after the engine had been stopped for a few minutes to make adjustments. They are direct evidence of the Hot Restart Problem. This data demonstrates the effect of the petrol boiling in the jet and weakening the mixture.

Carburettors are volumetric devices, ie for a given volume of air entering the engine, they add a fixed volume of petrol. When the petrol starts to boil, liquid and bubbles of vapour leave the jet. These are less dense than liquid on its own. As a result, a lower mass of petrol is delivered into the air stream resulting in a weak mixture. Weak running further increases the engine and engine bay temperatures, resulting in more fuel boiling. The mixture becomes so weak the engine splutters to a stop. This scenario is typical of stop-start driving in a queue of traffic.

On the XPAG engine, the carburettors are positioned above the hot exhaust manifold. This is the most obvious source of heating. There is a second, less obvious source: heat from the engine either conducted through the inlet manifold or hot gases from the cylinders. To identify the source of the heating, thermocouples were fitted to each carburettor:
- Air inlet – the temperature of the air flowing into the carburettor
- Transfer pipe ('Xfer pipe' on the graphs in Figure 7-5) – the connection between the float chamber and jet. This is the closest part of the carburettor to the exhaust manifold.
- Manifold – the connection between the carburettor and inlet manifold to measure heat flowing out from the engine.

The histograms in Figure 7-5 show the number and range of temperature readings for the six thermocouples (three on each carburettor). The red area corresponds to the temperatures where petrol vaporisation would cause problems.

While these do not reflect the temperatures that may be reached in an enclosed engine bay, they do show some interesting features.

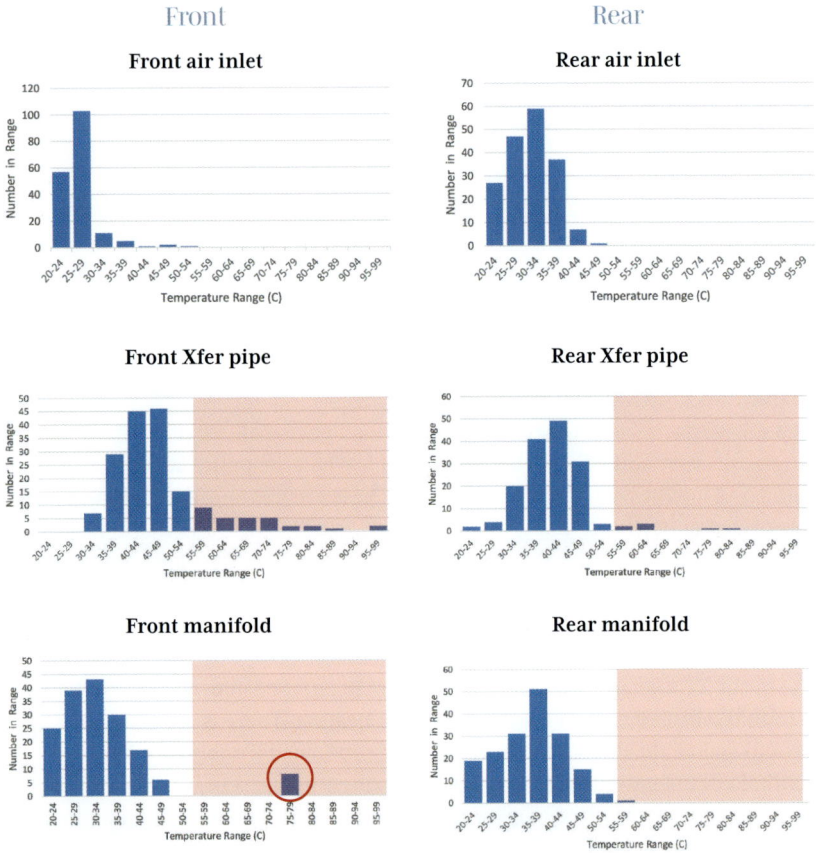

7-5: Carburettor temperatures.

The test engine was mounted on the dynamometer with no restrictions to airflow. Other than the exhaust manifold, there are no hot areas near the carburettors. The temperature of the air inlet of the front carburettor was lower than that of the rear (top graphs in Figure 7-5). This was due to the air being heated by the exhaust pipe which ran underneath the rear of the engine.

These two temperature profiles are reflected in the manifold temperatures (bottom graphs). When the engine is running, the temperature of the carburettor body reflects that of the inducted air. Where possible, the air inlet to the carburettors should be positioned to induct the coolest air possible, maybe installing ducting to direct cold air to the carburettors.

One set of readings stand out. The high front manifold reading, shown in the red circle (bottom left-hand graph on Figure 7-5). This corresponds to the five tests mentioned above, where the mixture was very weak.

Only the front carburettor was affected. It is probable that when the engine had stopped, the inlet valve on either cylinders one or two were open. This allowed hot gases from the cylinder into the inlet manifold raising its temperature. This observation is consistent with the heat soak tests, discussed in Chapter 9. These showed heat from the inlet manifold was the main factor in increasing the temperature of the carburettors after the engine had stopped.

When stopping a hot engine with the intention of restarting it a few minutes later (eg when filling up at a petrol station), it may be worth revving the engine and turning off the ignition while it is still running. Keeping the throttle open as the engine runs down will allow cold air into the cylinders, which may prevent this problem.

The transfer pipe temperatures (middle graphs in Figure 7-5) were sufficiently high in some test runs to cause issues with vaporisation. Other than the five cases highlighted above, a weak mixture was not observed in any other tests. It is probable the high fuel flow rate through the transfer pipe did not allow the petrol to vaporise before it reached the jet.

These measurements show that the mechanisms by which the fuel is heated are not what they may seem. The most obvious source of heating is the cast iron exhaust manifold positioned below the carburettors. The data shows it is heat from the cylinder head and hot exhaust gases returning up the inlet manifold that are the main source of heating.

Choice of petrol

Figure 7-4 shows the data for the 95RON (85MON) petrol.

Figure 7-6 compares the distillation curve for this petrol with that of a super grade petrol (red line). The 95RON (85MON) (blue line) is more volatile at the 75°C to 80°C temperatures seen in the front carburettor. At these temperatures, 45 per cent of 95RON (85MON) would have vaporised, compared with only 35 per cent of the super grade.

It is possible this effect would not have been as severe had the super grade petrol been used. This observation enforces the fact that no two brands or grades of petrol are the same. It is advisable to try using different brands and grades and use the one that causes your vehicle the fewest problems.

Percentage evaporation v/s temperature

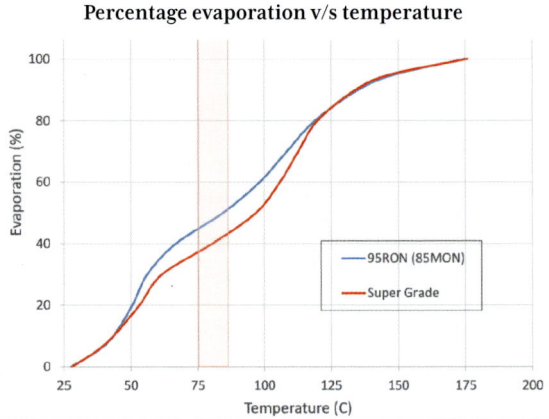

7-6: Volatility comparison 95RON (85MON) v/s Super Grade.

Summary

This data demonstrates the reasons for the Hot Restart and related problem that occurs in slow moving traffic are a direct result of the low temperature volatility of modern fuels.

The five tests, where the carburettors delivered a weak mixture, show the cause of the problem is not always the obvious one. Hot gases from an open inlet valve were probably the reason, rather than the heat from the exhaust manifold.

The transfer pipe temperatures are sufficiently high to cause vaporisation. While driving at normal road speeds, the volume of fuel flowing through this pipe prevents this happening. However, in slow moving traffic, much less fuel is flowing. Under these conditions, even a small temperature increase can cause vaporisation. The resulting weak mixture will stop the engine.

On the positive side, of the fuels tested, the super grade is less volatile over the measured temperature ranges, making the problem less severe.

Chapter 10 discusses the volatility of different fuels and the best choices to avoid the Hot Restart Problem.

8 Results from the tests – slow combustion

This chapter continues the presentation of the carburettor data from Chapter 7. These data showed an unexpected enrichment of the mixture at less than 3000rpm. This is referred to as the Enrichment Effect.

This finding is very significant. An engine usually runs at less than 3000rpm when driving on public roads. The tests suggest this Enrichment Effect is caused by a high degree of cyclic variability upsetting the operation of the carburettor. Chapter 6 describes why a high degree of cyclic variability can cause engine damage. Hence this finding highlights one of the main dangers of using modern fuel in classic engines. In contrast, the XPAG ran well on all the fuels at high rpm, full throttle, typical of its use when racing.

This chapter presents the carburettor data and the other evidence supporting this conclusion.

Unexpected results

It is worth reviewing Figure 7-3 in Chapter 7. This shows the expected air/fuel equivalence ratio (lambda) in relation to the carburettor suction piston height. Figure 8-1 shows the data from all the tests. The five weak data points, identified in Chapter 7, (two green and three red) can be ignored.

In Figure 8-1, the data points in the red circle show the carburettors delivering *different* lambda values for the *same* piston height. This is not

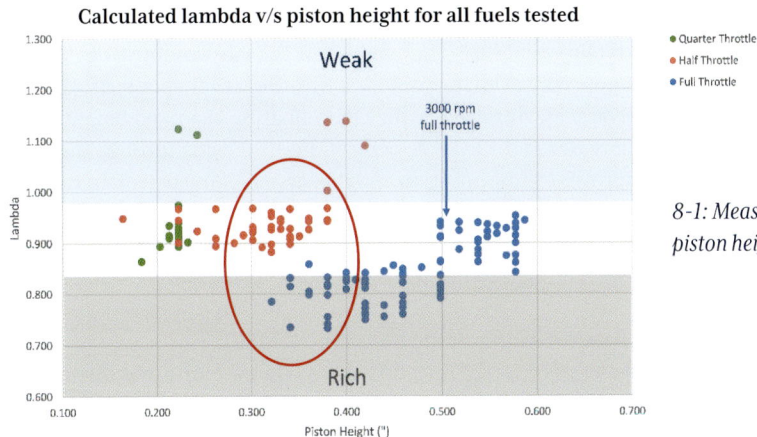

Calculated lambda v/s piston height for all fuels tested

8-1: Measured piston heights.

possible. The volume of fuel is fixed by the needle diameter at that piston height. The same piston height, the same volume of air, the same volume of fuel, hence the lambda values should also be the same.

At piston heights between 0.3in and 0.4in, the full throttle tests (blue dots) show a richer mixture than the half throttle tests (red dots). This is referred to as the Enrichment Effect. The only explanation for its cause is that something is affecting the operation of the carburettors.

Enrichment effect

The airflow through the carburettor is not constant, it is pulsed. Air is only drawn through when cylinders undergo the induction cycle. With a single carburettor, 4-cylinder engine, this is twice per revolution of the crankshaft. Once per revolution for 4-cylinder engines with twin carburettors. The inertia of the suction piston and the gases in the inlet manifold normally smooth out this pulsing airflow.

The inlet valve starts to open in advance of the piston reaching top dead centre (TDC). The exhaust valve does not close until the piston has passed TDC. On the XPAG, the 35° when both valves are open is called valve overlap. This is described in Chapter 4.

A high degree of cyclic variability can result in a delayed combustion cycle. The pressure in the cylinder will be high when the inlet valve opens, causing a pulse in the inlet manifold. This pulse will increase the pressure in the choke of the carburettor, causing the suction piston to fall before the induction stroke starts. When induction begins, both the inertia of this piston and the damper will slow the rate at which it rises. Under these conditions, the carburettor will deliver a richer mixture.

The Enrichment Effect seen in the full throttle tests below 3000rpm is caused by this high level of cyclic variability. They show how the late-combusting cycles affect the way the engine runs. During some full throttle, lower rpm tests, the suction piston could be seen vibrating, rather than floating at a fixed height.

As engine rpm increases so does the velocity of the air flowing through the carburettor. This improves the atomisation and dispersion of the petrol. In turn, this reduces the magnitude of the cyclic variability and size of the back pressure pulse. This is the reason why the mixture returns to normal as piston height (engine rpm) increases. Ultimately, delivering the correct lambda of around 0.95 above 3000rpm.

There are three possible causes of the Enrichment Effect:
- The characteristics of carburettors and the way they atomise and disperse the petrol into the inflowing air.

- The fuel that is being used.
- The design of the XPAG engine,
   specifically the valve overlap timing.
These points are discussed below.

Petrol atomisation

To improve petrol atomisation and
dispersion, tests were run using a nebuliser.
The Oxford English Dictionary defines a
nebuliser as a "device that converts a liquid
into a fine mist or spray." The nebuliser was
made from a thin nickel foil with eight-
micrometre holes (shown in Figure 8-2).
It was fitted into the gasket, between the
carburettor and inlet manifold.

As the air/fuel mixture passes through
the foil, the droplets of petrol are forced
to break up. They become the same size of

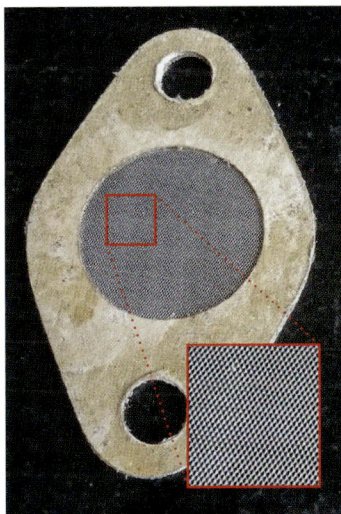

8-2: Nebuliser.

those produced by a fuel-injection system. The foil also increases turbulence
which, in turn, improves the mixing of the air and fuel. The net result is a
better dispersed mixture entering the cylinder, improving the cycle on cycle
combustion and producing a lower degree of cyclic variability.

The material of this thin foil is 30 per cent of its area, the remaining 70 per
cent is holes: it has a 70 per cent 'free area.' Although it looks solid in Figure 8-2,
in practice it caused very little restriction to the airflow.

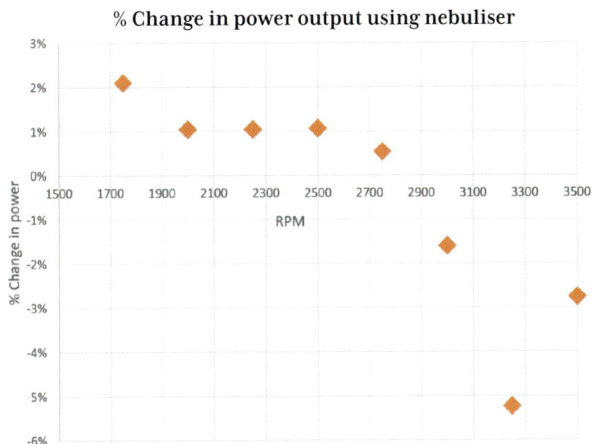

% **Change in power output using nebuliser**

8-3: Change in
power output.

Full throttle tests using the same 95 octane petrol and engine tuning gave different power outputs when the nebuliser was fitted. The change in power output from these tests is shown in Figure 8-3.

Below 3000rpm, the power output when using the nebuliser was increased by one per cent. Remember the simulation in Chapter 6? A lower degree of cyclic variability meant that more cycles were occurring at the optimum time, increasing power output.

Above 3000rpm the power output dropped. In this rpm range, the degree of cyclic variability was no longer affecting the way the engine ran. The nebuliser gave little benefit. The power drop was the result of the restriction in the inlet manifold created by the foil.

These tests show one of the causes of the Enrichment Effect is poor atomisation and dispersion of the petrol in the inflowing air stream, suggesting one mechanism to mitigate it. This is discussed in Chapter 11.

Unfortunately, a foil such as the one used in this test is not suitable for road use. Unless the air and petrol entering the engine can be filtered to remove any particles greater than eight micrometres in diameter, such a foil would soon block and choke the engine. A wire mesh is not a practical option for two reasons. Firstly, the holes are much bigger and will not cause the droplets of petrol to break up to the same extent. Secondly, meshes only have a 30 to 40 per cent 'free area.' This will significantly restrict the airflow into the engine.

Different fuels
Unburned hydrocarbons

As was discussed in Chapter 4, unburned hydrocarbons (HC) in the exhaust are a sign of how well the petrol burns in the cylinder. As the engine was fully tuned for each test run, it is possible to directly compare the HC data for each test. Remember the Enrichment Effect is the result of calculating the value of lambda had the engine not been retuned. Figure 8-4 shows the parts per million (ppm) of unburned hydrocarbons (HC) in the exhaust for four sets of tests.

- 95RON (85MON) petrol
- The same 95RON (85MON) petrol with 20 per cent added kerosene.
- The same 95RON (85MON) petrol with the nebuliser fitted (as described above)
- Petrol containing 10 per cent ethanol (E10)

The HC levels for the three tests using the 95RON (85MON) petrol are mostly the same. Below 3000rpm, the kerosene and nebuliser reduce the level of emissions, but only by a small amount. The large difference is when E10 is used. The differences mostly disappear above 3000rpm where the Enrichment Effect does not present a problem.

E10 contains oxygen linked to the hydrocarbons. By its nature, the oxygen is better dispersed in the mixture. As with the nebuliser, this improves the combustion and reduces the level of cyclic variability – one reason why it produces less HC than unblended fuels.

These results allay some people's fears that added kerosene does not burn properly and may cause engine damage. Over the whole rpm range, the petrol with the added kerosene produced very similar HC emissions as the standard petrol. It was burning just as well.

The differences in performance of the fuels that were tested is presented more fully in Chapter 10.

Full throttle unburned hydrocarbons

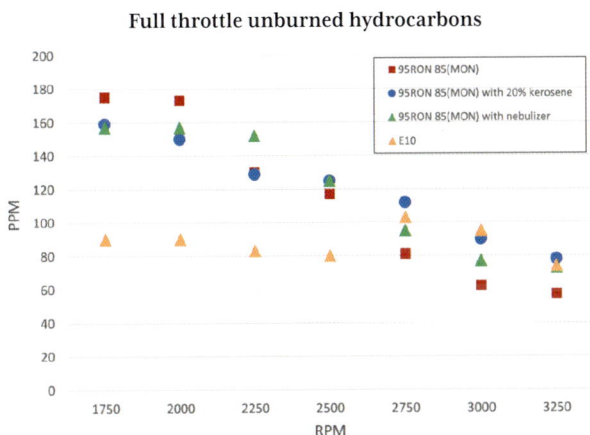

8-4: Unburned hydrocarbons.

Carbon monoxide

Carbon monoxide (CO) in the exhaust gas also gives an indication of how well the fuel burns. The higher the level of CO the worse the mixing and combustion.

The CO data from the four tests above is shown in Figure 8-5. The interesting part is the left-hand side of the graph below 2750rpm. Here, the percentage of CO in the exhaust gas differs by a factor of over two. Significant as the red line is for the 95RON (85MON) petrol and the blue line is the *same* petrol with 20 per cent added kerosene.

The results shown in Figures 8-4, 8-5 and 8-6 present a consistent picture. Below 3000rpm, the degree of cyclic variability depends on what fuel is being used. A high degree of cyclic variability affects the operation of the carburettor and the exhaust emissions. It increases the Enrichment Effect below 3000rpm and increases the levels of CO in the exhaust.

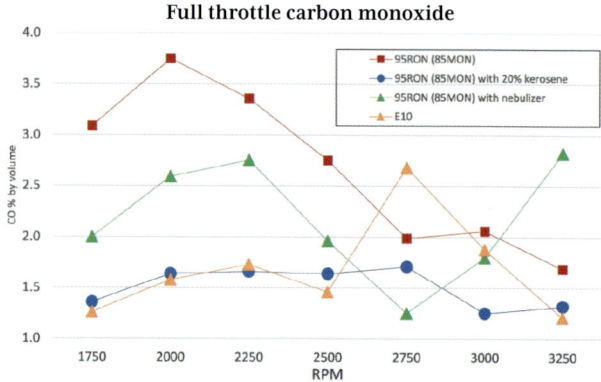

Full throttle carbon monoxide

Legend:
- 95RON (85MON)
- 95RON (85MON) with 20% kerosene
- 95RON (85MON) with nebulizer
- E10

8-5: Effect of different fuels on exhaust CO.

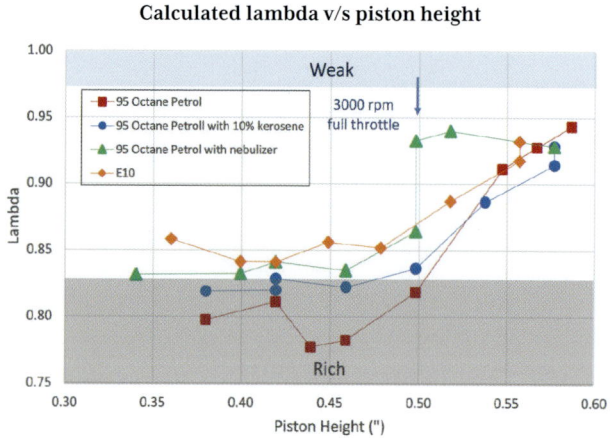

Calculated lambda v/s piston height

Legend:
- 95 Octane Petrol
- 95 Octane Petroll with 10% kerosene
- 95 Octane Petrol with nebulizer
- E10

Weak

3000 rpm full throttle

Rich

8-6: Different fuels.

Design of the XPAG engine

There are many factors that affect the combustion process. These include the design of the inlet manifold, and the type of carburettor and degree of valve overlap. Although the design of the XPAG is typical of many engines, it raises the question, are engines from other manufacturers affected in the same way?

The Enrichment Effect is caused by the way SU or variable jet carburettors work. Although it is unlikely that fixed jet carburettors will behave in the same way, these engines will likely still suffer from a high level of cyclic variability like the XPAG.

For each of these tests, the engine and its state of tune were the same. The only difference was the brand and grade of fuel being used. These changed the way the XPAG ran. More specifically, they appeared to alter the degree of cyclic variability. As a result, similar problems will likely affect engines from other manufacturers, but it is possible the rpm range where they are at their worst may be different.

Needle profile

One aim of the Manchester tests was to assess if the standard carburettor needle was suitable for use with modern fuels. The Enrichment Effect makes this difficult. Figure 8-7 shows the results for all the tests excluding the full throttle data below 3000rpm and the five weak tests discussed in Chapter 7.

The data shows that the original manufacturer ES needle is still a good choice for the XPAG engine. It delivers the correct air/fuel mixture over a wide range of operating conditions when using a variety of modern fuels. This includes the petrol containing 10 per cent ethanol, demonstrating the enleanment caused by ethanol is not problematic up to this concentration.

8-7: Standard needle profile.

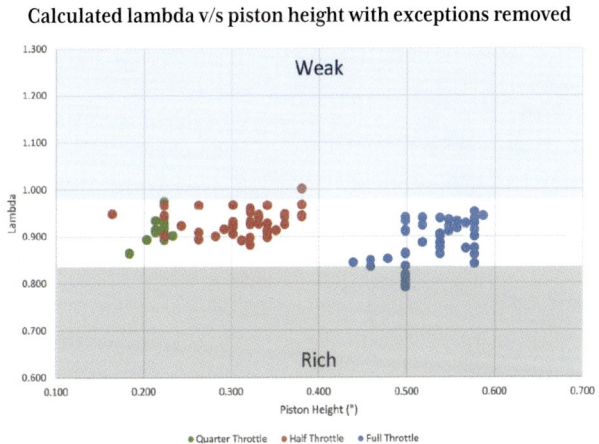

Calculated lambda v/s piston height with exceptions removed

Summary

What is of particular concern is that cyclic variability appears to be at its worst when driving on public roads. Not only does this cause damage to an engine, it also increases engine bay temperatures, making the Hot Restart Problem worse. While there is no magic solution, the Manchester tests identified a number of steps owners can take to mitigate this problem.

Chapters 9 to 13 suggest ways to improve the way the engine runs, especially for use when driving on the public highway. Some of them challenge conventional thinking about tuning engines.

The recommendations are summarised in Chapter 15.

9 Keep the fuel system cool

Compared to classic petrol, modern petrol is more volatile at temperatures below 50°C (122°F) – see Chapter 2. When a vehicle stops, there is no air flowing through the engine bay, petrol through the fuel system or water through the cooling system. Heat from the engine raises the temperature of the engine bay. It can easily reach the point where the fuel will start to boil in the carburettor.

As it boils, petrol blows out of the jet, running into the inlet manifold. The vapour bubbles prevent the carburettor from delivering the correct mixture. Try to start the engine and it coughs into life, running rich as the liquid fuel in the inlet manifold enters the cylinder. The engine will then stop because of the weak mixture. Sometimes, it is possible to restart the engine using the choke. After a few coughs and splutters, normal running returns as cold petrol fills the fuel system.

The reason engines stop in slow moving or stop-start traffic is similar. Even though the engine is lightly loaded and not generating much heat, there is little airflow through the engine bay. Engine bay temperatures rise. The reduced flow of fuel through the petrol pump, petrol pipes and float chambers gives more time for it to heat up and boil. Vehicles run without problems in moving traffic, misfiring and stopping when the traffic flow drops to a near standstill.

These problems can be avoided by keeping the fuel system components as cool as possible. Especially those in the engine bay such as: the carburettor(s); fuel filter (if fitted); fuel pump and fuel lines. This chapter discusses ways this can be done.

Warning: Classic vehicles are not the same. Differences will have accumulated with the passage of time. Also, the severity of problems varies immensely, even between the same models. A solution that may work well for one vehicle may not resolve the same problems in another. The suggestions should be taken as just that; they are not intended as solutions to be blindly adopted.

Where to start
One of the findings from Manchester was that the sources of heat were not always the obvious ones. For example, even though the carburettors on the XPAG are positioned directly above the hot exhaust manifold, the most obvious source of heat, after the engine had stopped, hot gases from the cylinders in the inlet manifold appeared to cause the greatest problem.

This highlights the importance of ensuring any steps to mitigate problems are investigated before implementing them.

It is worth buying an infrared thermometer and a multimeter with a thermocouple. A thermal imaging camera is the ideal way to identify high temperature areas. While this is the most effective solution these are expensive. Suppliers can be found by searching the internet.

Furthermore, take care when investigating engine bay temperatures. Stopping the vehicle and opening the bonnet (hood) will allow the hot air to escape, changing the temperature profile.

Consider using a multimeter and thermocouple. This allows temperatures to be measured without opening the bonnet. Extend the thermocouple wire into the cockpit. This allows a passenger to check temperatures while the vehicle is being driven.

Steps to consider

There are many steps an owner can take to help keep the engine bay and fuel temperatures as low as possible. These are:

- Keep the engine cool.
- Reduce the amount of heat reaching the fuel system.
- Change fuel pipe runs.
- Reduce the volume fuel in the engine bay.
- Fit electric fan(s).
- Insulate the fuel system.
- Install heat shields and thermal spaces.
- Add baffles.
- Insulate the exhaust system.
- Use of gears.
- Think about where you park.

Keep the engine cool

As 35 per cent of waste heat from the engine is lost through the cooling system, it is important to make sure this is working efficiently. For water-cooled engines:

- Flush out the radiator.
- Remove insects and other debris from the radiator fins.
- Check the water hoses are in good condition and not blocked.
- If fitted, ensure the thermostat is working properly.
- Once every two years or so, flush the engine and radiator with clean water to remove the build-up of any debris.
- Use soft or distilled water to refill the cooling system.

- Use a cooling system wetting agent. It is claimed these help to reduce temperatures.
- Check your fan is fitted the correct way around. If the blades are dished, the convex face (outward bulge) should face the radiator.
- Ensure the fan belt is in good condition, is driving the fan properly and not slipping.
- Ensure air can flow freely through the engine compartment.
- For air-cooled engines, check for and remove any debris from the cooling fins around the piston bore and cylinder head. Make sure the fins are clean.
- Remove leaves and other blockages such as badges from air intake grilles.
- Ensure ancillary equipment, such as the horn, wiring, etc, are not blocking the airflow. This is particularly important around fuel pipes, filters, carburettors, and the fuel pump (if it is located in the engine bay).

Hot air rises. Areas around the top of the engine bay will be hotter than those at the bottom. Any fans, ducts or baffles intended to move cooler air through the engine bay should encourage this natural flow, not 'fight it.' Blow cold air in at the bottom of the engine compartment, extract hot air from the top.

Change fuel pipe runs

Ensure the fuel system components and air inlet are not placed near hot parts of the engine. If possible, move them into the cool airflow. Position them as low as possible in the engine bay.

The layout of the fuel system on an MG TC is far from ideal.

The petrol pump and pipe linking it to the carburettors (circled on the right) are in the hot air at the top of the engine bay. The air intake (circled on the left) also draws in this hot air.

Suggestions to improve this arrangement would be to run the braided pipe from

9-1: MG TC. Example of poor positioning of fuel system components.

the bottom of the fuel pump rather than from the top. Pancake air filters fitted to the carburettor would draw in colder air from the slits in the bonnet (hood).

Another example is shown in Figures 9-1a and 9-1b.

Figure 9-1a shows the standard routing of the pipe linking the two carburettors on the MG TC. The braided pipe is just visible passing underneath the black air filter. It passes directly over the top of the exhaust manifold.

Figure 9-1b shows the modification. The pipe is looped over the top of the air filter away from the exhaust manifold into the cooler air.

Some models of Triumph cars suffer from similar issues with the routing of their fuel pipes. In this case, the pipe from the fuel pump to the carburettors runs around the front of the engine. Passing behind the radiator cooling fan. Hot air from the radiator is blown directly onto this pipe. Re-routing the pipe to pass below the radiator fan may be one option.

9-1a : MG TC before re-routing ...

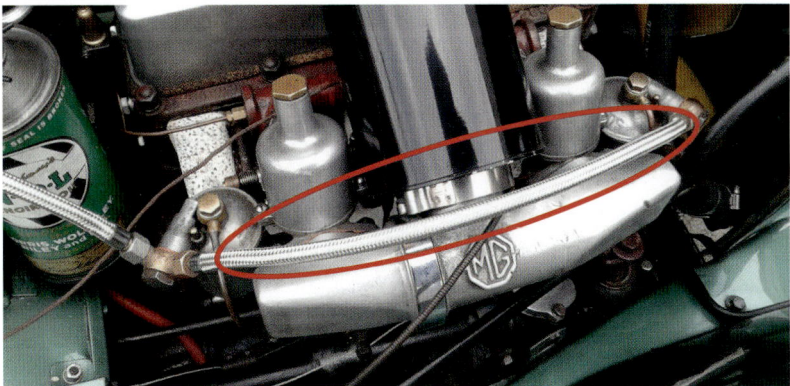

9-1b: ... and after re-routing.

Another example is the type and position of an electric fuel pump. On the MG T series cars, the fuel pump on later models was changed from a low pressure pump fitted in the engine bay to a high pressure pump fitted at the rear of the car. The reason for this is not clear. When pumping a liquid it is always better to 'push it' rather than 'suck it.' This avoids cavitation, especially important where the pipe from the petrol tank to the pump is exposed to hot areas, for example around the exhaust system.

If the vehicle is fitted with an electric fuel pump, it may be worth repositioning it nearer to the petrol tank. This will reduce the problem of cavitation. Also, the higher pressure in the fuel lines between the pump and carburettor will increase the petrol's boiling point.

Warning, extreme care needs to be taken when positioning fuel system components. The top of the petrol feed to the carburettor must be higher than the top of the petrol tank. If not, petrol can still flow under gravity after the engine has been switched off. The early MGs are fitted with a large, upright slab tank at the rear of the car. In these cars lowering the fuel pump risks dangerous petrol leaks from a full tank.

Reduce the volume fuel in the engine bay

Try to minimise the volume of fuel in the fuel system components situated in the engine bay. The less hot fuel there is, the less vapour will be generated. It will cool down more quickly and will take less time for cooler petrol to reach the engine.

Keep fuel pipe runs short. Avoid 'reservoirs' such as large fuel filters, etc. The example shown in Figure 9-1b uses petrol pipes with a narrower bore. The replacement pipes have a ¼in (6mm) bore rather than the original ⁵⁄₁₆in (8mm) bore. This both reduces the volume of petrol and causes it to flow through the pipes faster, giving it less time to get hot.

Repositioning the fuel pump, as described above, offers another suggestion.

This creates a dilemma. Is it better to have a longer fuel pipe passing through cooler areas of the engine bay or a shorter one that holds less petrol? This is one reason why it is necessary to identify the hot spots and adopt the most appropriate solutions.

Fit electric fan(s)

Electric fans fitted to the radiator help to keep air circulating, particularly in slow moving traffic. Unfortunately, they may make matters worse. The fan draws cool air through the radiator where it is heated, possibly to more than 90°C (200°F). It blows this hot air into the engine bay – not ideal. It may be better to position the fan at the bottom of the radiator. This allows it to suck in cooler air.

Fit a timer or similar circuit to keep the fan running for five to ten minutes after the engine has stopped. This will help disperse the hot air from the engine bay. Another possibility would be to add a switch or circuit to reverse the voltage polarity to the fan. This would draw cool air from under the car and vent the hot air through the front of the radiator. It could be used when the vehicle is stopped or moving slowly.

As well as a radiator fan, another suggestion is to fit 12-volt computer fans to blow cold air onto critical fuel system components.

Insulate the fuel system

There are insulating products available on the market that are suitable for use in an engine bay. They both reflect infrared radiation and protect against conducted heat.

Unfortunately, insulation does not stop the transfer of heat, it only slows it down. Once the engine has stopped and the petrol is no longer flowing, it will heat up, no matter how well insulated. Benefits will only arise if this heating is delayed long enough to allow the engine bay temperature to fall below 45°C (113°F).

As was found at Manchester, it is not always obvious where heat is coming from. After stopping the engine, temperature readings from the carburettors were recorded every 30 seconds.

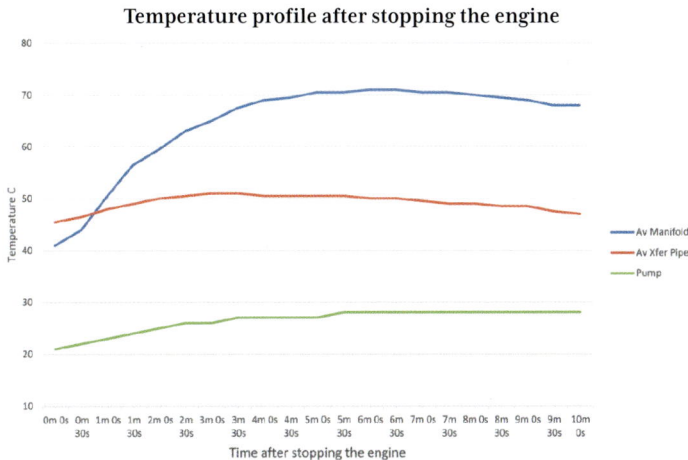

9-2: Temperature profile after the engine was stopped.

These are shown in Figure 9-2, which shows: temperature of the fuel pump (green); average temperature of the two transfer pipes between the float chamber and carburettor (red); and average temperature of the junction between the inlet manifold and carburettor (blue).

The greatest temperature rise was at the junction between the inlet manifold and carburettor. This is due to heat conduction from the 170°C (340°F) cylinder head and the hot gases travelling back up the inlet from the valves. After four minutes, the temperature of the carburettor body was 70°C (160°F). At this temperature, 50 per cent of modern petrol will evaporate, hot enough to cause the Hot Restart Problem.

In Figure 9-3, a normal picture is shown alongside a thermal Figure. This is taken looking down the inlet tract with an open throttle. The suction piston has been lifted with a screwdriver. The light blue area within the circle is the carburettor needle. This is at 61.1°C (142°F). The green colour around it shows the higher temperature gases in the inlet manifold. The bottom part of the carburettor and jet are much cooler (darker blue in colour), even though they are close to the very hot exhaust manifold – rust coloured on the normal picture and yellow, red and white on the thermal image.

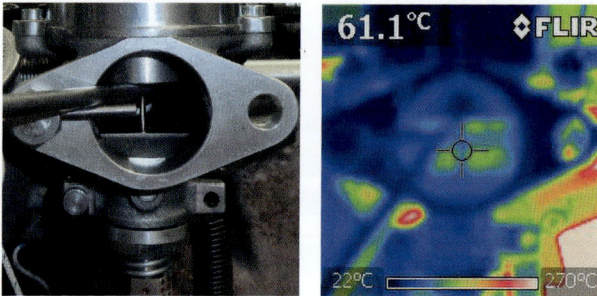

9-3: Heat from the inlet manifold.

Figure 9-3 reinforces the source of the temperature rise seen in Figure 9-2. The main source of heat is directly from the engine, not through radiation from the hot exhaust manifold. In this situation, fitting a heat shield may not help. Thermal spaces between the carburettor and inlet manifold may be more effective.

When you stop an engine, it is likely one cylinder will have its inlet valve open. This will allow very hot gases to leave the cylinder and flow back into the inlet manifold where they will heat the carburettors. One possible way to address this may be to blip the throttle and switch off the ignition while the engine is still running. Opening the accelerator as the engine runs down will allow cold air to enter the cylinders, venting the hot gases through the exhaust.

Figure 9-4 is a thermal image taken when the engine is running at full throttle. The white, very hot, exhaust manifold is contrasted against the cold, dark blue, carburettors.

The cross fuel pipe that links the two carburettors is also cold.

Although the engine is producing a colossal amount of heat, the volume of petrol flowing through the carburettors and cross fuel pipe is sufficient to keep them cold.

The carburettors are also cooled by the small percentage of the petrol that evaporates as it leaves the jet. This is the darker blue band seen in the inlet of each carburettor. The cooling effect is particularly evident with the suction chambers that are less than 20°C (68°F).

The only parts of the carburettors that are relatively hot are the choke levers, shown in yellow. Insulating these may be something worth considering.

In slow moving traffic, less fuel is being used. The flow rate to the carburettors decreases as does the cooling effect from evaporating petrol. The petrol gets hotter and starts to boil, weakening the mixture and causing the engine to stop.

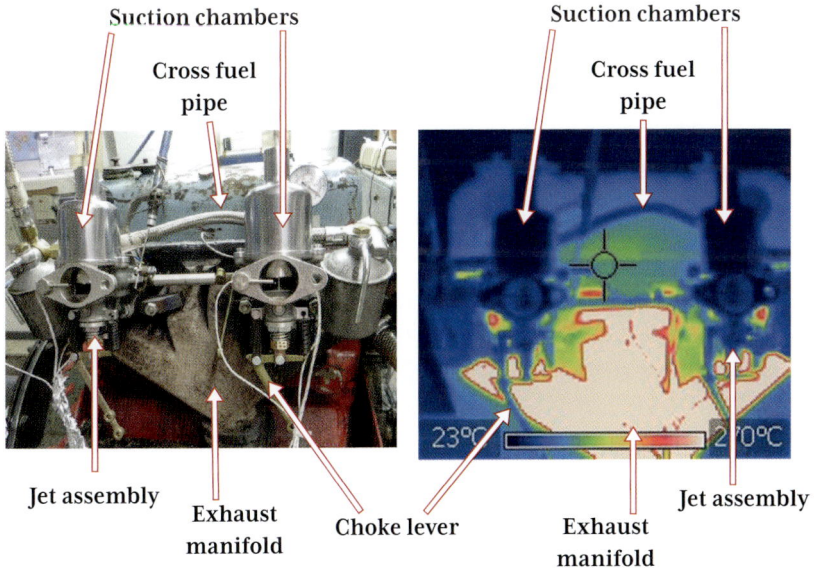

9-4: Engine running at full power.

Install heat shields and thermal spaces

A heat shield between the exahust manifold and carburettor(s) stops heat radiated from the exhaust manifold reaching the carburettor(s). Thermal spacers between the carburettor and inlet manifold protect against heat conducted from the engine. In a similar way to insulation, these measures only slow the transfer of heat. Their effectiveness will depend on how well the carburettors are cooled.

If a heat shield is already fitted, consider adding aluminised insulation on the side facing the exhaust manifold.

Add baffles

On some engines, it may not be possible to move fuel system components away from hot areas. In this case adding baffles may offer a solution. These can be made from thin steel or aluminium and fitted using existing bolts.

Baffles can serve one of two purposes. They can either redirect hot air away from fuel system components or direct cooler air onto them. It is better to direct cold air onto the components rather than redirecting hot air away. The difficulty is knowing which way the air is flowing through the engine bay when the car is moving. Try to duct cool air from the front of the vehicle, before it has passed through the radiator. Alternatively duct air from the bottom of the engine bay where it is cooler. Unfortunately, fitting baffles may not be easy.

While baffles will help to keep fuel system components cooler when the vehicle is moving, they are not effective when it has stopped, as there is no airflow through the engine bay, unless an electric fan with a run-on facility has been fitted.

Warning: if baffles are fitted, take care they do not rub against and damage wiring or fuel pipes. Thin steel or aluminium sheets can have very sharp edges after they have been cut. They may also vibrate when the engine is running if they are not braced.

Insulate the exhaust system

As can be seen from Figure 9-4 the exhaust manifold and downpipe(s) from the manifold to the exhaust are very hot. These both radiate and convect a great deal of heat in the engine bay. Insulating them will reduce their effect. In particular, consider insulating the exhaust manifold and downpipe where they are close to the fuel system components.

As with insulating the fuel system components, the insulation will only slow, not stop, the flow of heat. Once the vehicle has stopped moving, insulation on the exhaust may not provide an effective solution.

Use of gears

A spark-ignition engine, such as the XPAG, is around 30 per cent efficient. It only converts about a third of the heat energy in the petrol into power. The remaining two thirds of the heat goes into heating the engine, exhaust and cooling system. Driving in the rpm/load range where the engine runs most efficiently not only saves petrol, it also helps reduce engine bay temperatures.

The suction piston height in the SU carburettors provides the means to assess how much petrol is flowing into the engine.

The calculation of the exact volume of fuel is complex. However, for one fuel it allows comparisons to be made.

Knowing how much petrol is flowing into the engine and its power output enables the engine's relative efficiency to be calculated.

This is shown in Figure 9-5 for different throttle settings and rpm.

The quarter and half throttle efficiencies (blue and green lines) are what would be expected: constant up to 2500rpm and 3000rpm after which they drop. This is caused by the throttle butterfly restricting the flow of mixture into the engine at higher rpm.

In contrast, below 2500rpm, the full throttle efficiency (red line) is on average 7.5 per cent lower than that at half throttle. This is caused by a high degree of cyclic variability reducing the power output. For normal driving, the full throttle efficiency would be even lower below 3000rpm, due to the Enrichment Effect.

Relative % efficiency (Super Grade)

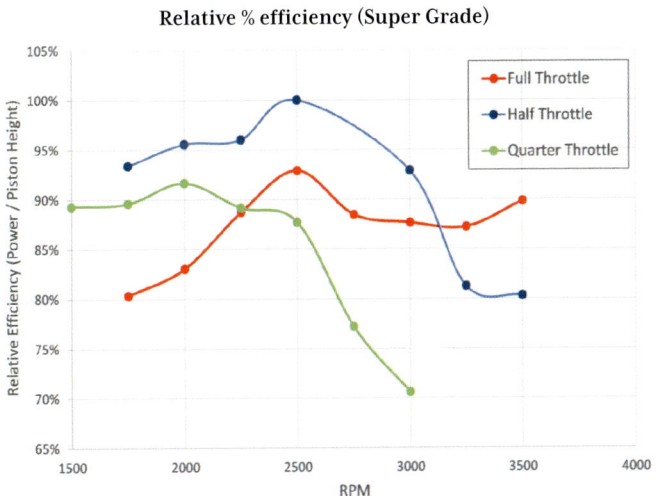

9-5: Relative efficiency.

To ensure the engine is running at its most efficient, use the gears. Avoid the use of full throttle at low rpm, change down, and do not let the engine rev on part throttle, change up!

Think about where you park

If you leave your vehicle in the sun on a hot day, it is most likely that not only the engine bay, but also the fuel in the tank will get hot. This is particularly true of older cars or motorbikes with exposed petrol tanks.

This will cause the front end components to evaporate, resulting in the petrol 'going off' as discussed in Chapter 2. When you start to drive your vehicle the already hot petrol will enter the carburettors. This will increase the likelihood of it evaporating and causing the engine to stop.

Try to park in the shade or cover the petrol tank with an aluminised sheet to help keep it cool.

Summary

The volatility of modern petrol at relatively low temperatures is the ultimate cause of the Hot Restart Problem and engines stopping in slow moving traffic, especially on hot days. This chapter has suggested several ways to keep the fuel cool to help mitigate these problems. Unfortunately, there is no simple magic solution.

10 Choice of fuel

Chapter 2 showed the volatility profiles of different brands and grades of petrol were not the same. Chapter 8 showed that the way the XPAG engine ran depended on the fuel it was using, particularly below 3000rpm. This data shows the composition of modern petrol from different sources is anything but the same.

Not only does each brand use different additives, many retailers also sell different grades. These include standard grade 95RON (85MON) and super grade 97RON (86MON). To further complicate matters, the composition of petrol can vary by geographical region and over the time of year.

For example, the United Kingdom's fuel distribution industry is served by six main refineries. These supply petrol to their local area. Many of these came on stream in the late 1950s and early 1960s, reflecting the postwar demand for petroleum. No two refineries are identical. The composition of a brand of petrol most possibly changes depending on where in the UK it is purchased.

Additionally, in the UK and some other countries, different types of fuel are sold throughout the year. For example:
- Winter fuel (October to April in the Northern Hemisphere): This has the highest percentage of front end components to make it easier to start a cold engine. Winter grade fuel was used in the Manchester tests.
- Intermediate fuel (April to May and September to October).
- Summer fuel (June to August). This has the lowest percentage of front end components and is the best fuel to use in a classic vehicle.

Unfortunately, these dates are not fixed. They vary with ambient temperature and turnover at any particular filling station. Making it difficult to know what type of fuel is being sold.

Chapter 8 showed the engine suffered from a high degree of cyclic variability at typical engine rpm and loads used for driving on public roads. It also showed how this effect could be reduced by the choice of fuel. This chapter reviews the fuels tested at Manchester and ranks their performance in reducing the degree of cyclic variability.

Fuels tested
The fuels that were tested include specialist fuels, those containing ethanol, normal and super grade fuels. Specifically:

Specialist fuels:
- Sunoco Optima 98 – Ethanol free fuel with a long storage lifetime.
- Sunoco 101 – 101RON (90MON) a high octane racing fuel containing approximately seven per cent ethanol.
- 50 per cent Toluene – Toluene is already used in many fuel blends but not at this concentration. Toluene is an octane booster that does not dilute the energy content of the fuel. It vaporizes easily.
- Avgas – Used in light aircraft. Like petrol from the 1960s, it uses tetraethyl lead as an octane booster.

Commercially available fuels:
- 95RON (85MON) – Containing approximately five per cent ethanol. This is sold as standard grade fuel.
- 95RON (85MON) – Ethanol free. Again sold as standard grade
- Super grade (with ethanol) – 99RON (86MON) containing approximately five per cent ethanol. Sold as a super grade petrol.
- Super grade (ethanol free) – 97RON (86MON), a different brand of super grade petrol.
- E10 – 95RON (85MON) standard grade of petrol sold on continental Europe with a ten per cent ethanol content.

Additional tests:
- 95RON (85MON) – Ethanol free petrol. Nebulisers were fitted between the carburettors and inlet manifold. These improved atomisation and dispersion of the petrol at the expense of restricting the flow of mixture into the engine.
- 95RON (85MON) – Ethanol free petrol with 20 per cent added kerosene by volume. This percentage is double that normally suggested. It allowed the extreme effects of adding kerosene to be investigated.

Avoiding the Hot Restart Problem

Chapter 2 described how the volatility of modern petrol causes the Hot Restart Problem. Fuels with a low percentage of front end components directly address this problem. A smaller volume of these fuels will evaporate at engine bay temperatures.

WARNING: Specialist equipment is needed to measure the volatility profile of a fuel. It is not something that should be attempted at home. Petrol vapour is highly flammable. **Under no circumstances should an attempt be made to measure how much evaporates at a given temperature.**

The volatility profiles of some of the fuels tested at Manchester were measured in a professional laboratory. These are summarised in Figure 10-1. There are significant differences, especially at the temperatures experienced in the engine bay.

A petrol with a lower volatility in the 45°C (113°F) to 70°C (160°F) range will result in less petrol evaporating at typical engine bay temperatures.

Volatility profiles for all fuels tested

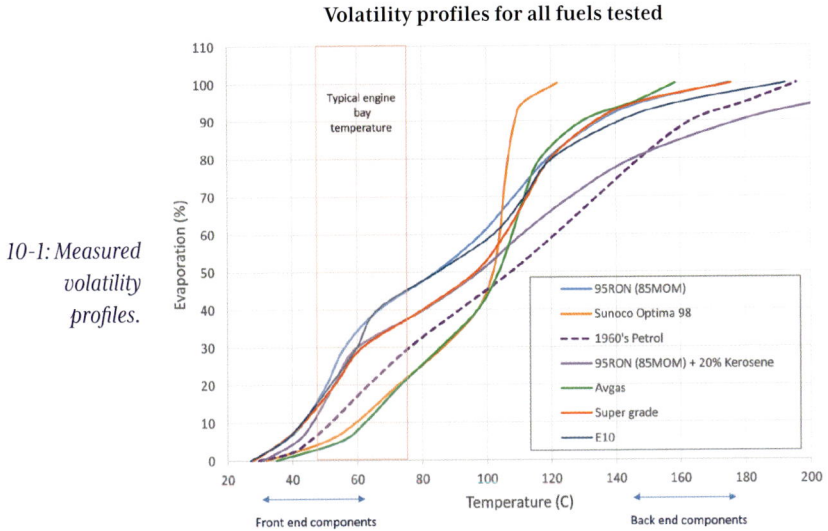

10-1: Measured volatility profiles.

What fuel to use to avoid the Hot Restart Problem?

Figure 10-2 ranks the fuels by the volume evaporating at 50°C (122°F). Specialist fuels are shown in grey, those containing ethanol in orange.

The two best fuels are Avgas and Sunoco Optima 98. Both these have fewer front end components than commercially available petrol. Their volatility profile matches that of the 1960s fuel below 100°C (212°F).

The use of either of these fuels will resolve the Hot Restart Problem. Unfortunately, it is not legal to use Avgas in a road vehicle. While Sunoco Optima 98 is expensive, it could be considered as the fuel of choice for low mileage vehicles. See the section at the end of this chapter.

A more practical solution is to use a super grade fuel, or add kerosene (paraffin) (see section at the end of this chapter). These do not reduce the front end components as much as Sunoco Optima 98 or Avgas. However, a smaller fraction will evaporate at engine bay temperatures than standard petrol.

Adding 20 per cent kerosene to the 95RON (85MON) petrol reduces its

Percentage evaporation at 50°

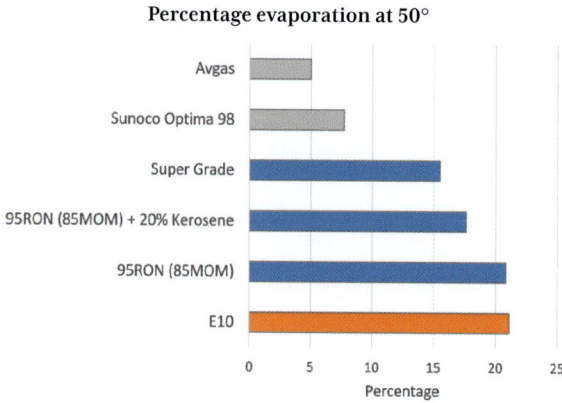

10-2: Percentage evaporation at 50°C.

volatility below 50°C. A better solution is to add kerosene to a super grade petrol to further reduce its low temperature volatility.

Distillation tests must be performed by specialists. To find a fuel's specification visit https://classicenginesmodernfuel.org.uk/bestfuel/ where owners can rank the brands and grades of fuel that best suit their vehicles. Alternatively, perform an internet search for a fuel brand and type followed by 'product data sheet.' This may give a reference to the manufacturer's test data for that fuel.

Typically they will contain a Distillation section, eg:

Parameter (BS Methods)	Units	BS or HM Revenue & Customs Limit		Typical
		Min	**Max**	
...				
Distillation:				
Evaporated @ 70°C	% (V/V)	22.0 (W) 20.0 (S)	50.0 (W).48.0 (S)	**42 (W) 40 (S)**
Evaporated @100	% (V/V)	46.0	71.0	59
Evaporated @ 150	% (V/V)	75.0	-	91
Final Boiling Point	°C	-	210	185
Residue	% (V/V)	-	2	1

The figure of interest is the 'Evaporated @ 70°C.' For the winter version of this fuel (W), 42 per cent of its volume will have evaporated at 70°C, for the summer fuel (S) this figure is 40 per cent. Choose a brand and type of fuel with the lowest figures for the evaporation at 70°C.

Regardless of which fuel you use, it is best to avoid filling your tank on the first run of the season. Petrol stations are probably selling the more volatile winter fuel. Only put in the fuel you need and fill up as soon as summer fuel becomes available. Otherwise, you could end up with a tank full of volatile winter or transition petrol on a hot summer's day.

Fuels for the racers

The findings described in Chapter 8 is significant. They show there is a difference in the way the engine performs above and below 3000rpm.

For example, when raced, the engine will mostly run in excess of 3000rpm on full throttle.

Figure 10-3 ranks the average full throttle torque between 3000rpm and 3750rpm. These are shown as a percentage reduction from the best performing fuel. At Manchester, concerns over damaging the engine restricted the maximum to 3750rpm.

The difference between the best and worst performing fuels is small at six per cent. However, there is still a difference. There appears to be no significant differences between the standard or high octane fuels or those that contain ethanol.

As a result, it is not possible to advise those who use their vehicles for racing.

Average torque – 3000-3500rpm

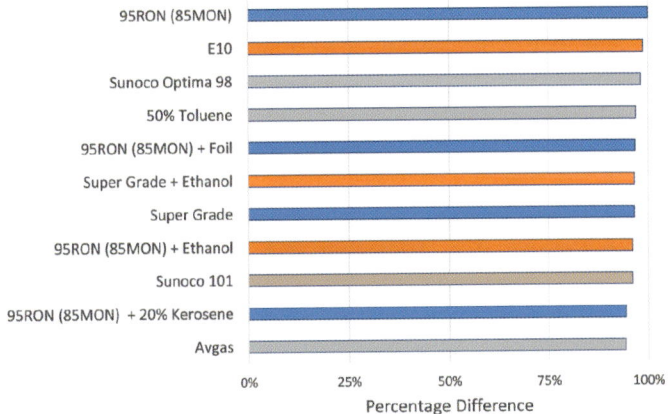

10-3: Average torque.

Fuels for road users

In direct contrast to racing, for road use the engine is normally run at less than

3000rpm on part throttle. Under these operating conditions, the degree of cyclic variability has a significant effect on the way the engine runs.

Chapter 8 described how the degree of cyclic variability depended on which fuel was being used. A fuel that reduces this degree will not directly address the Hot Restart Problem. It will help reduce possible damage to the engine and the temperatures in the engine bay, and these lower temperatures may reduce the severity of the Hot Restart Problem.

The average performance of the different fuels between 2000rpm and 3000rpm are ranked for the following:

- Carbon monoxide (CO) – High levels of CO in the exhaust gases is a symptom of poor combustion, an indicator of a high degree of cyclic variability.
- Unburned hydrocarbons (HC) – These are normally an indicator of the engine running too rich. As it was tuned for each test run, high levels of HC in the exhaust gases are a symptom of poor mixing.
- Relative efficiency – Chapter 9 described how the engine's efficiency varied. It suggested that high levels of cyclic variability reduced the efficiency below 3000rpm on full throttle. This reduction in efficiency also depended on which fuel was being used.

These three factors all depend on the engine's state of tune. Under normal circumstances it would not be possible to make comparisons. Changing the fuel may change the state of tune. At Manchester both the air/fuel ratio and ignition timing were set to the optimum for each test. Because of this, it is possible to make direct comparisons between these factors for the different fuels that were tested.

CO average – 2000-3000rpm

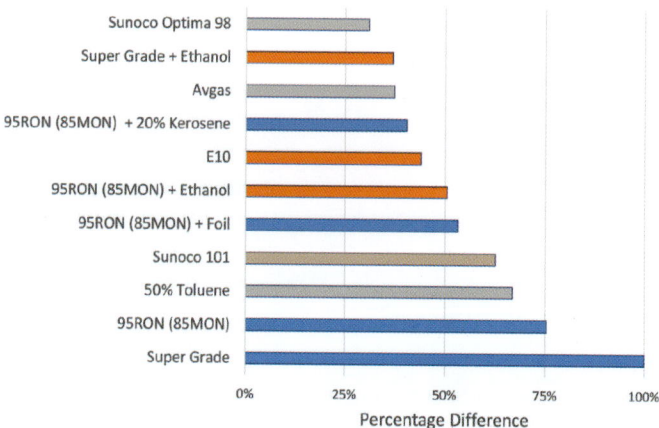

10-4: Average CO emissions.

Carbon monoxide (CO)

Figure 10-4 shows the average CO levels over the 2000-3000rpm range compared as a percentage of the worst performing fuel.

High levels of CO are a direct indicator of poor combustion caused by imperfect mixing of the air and fuel in the cylinder. The lower the level of CO, the better that sample of fuel is burning.

The best performing fuels produced under 50 per cent of the CO emissions than those of the worst. The grey bars show the specialist fuels, the orange bars or bars with orange shading showing the fuels that contained ethanol.

Unburned hydrocarbons (HC)

Like CO, high levels of unburned HC in the exhaust gases is an indication of poor mixing. The processes that lead to high levels of HC can be different from the production of CO.

Unless the mixture is rich (ie short of oxygen) all the volatile hydrocarbon molecules will disassociate and burn to H_2O, CO_2 and CO. If the combustion cycle is delayed, for example by cyclic variability, it is possible for some hydrocarbons to remain unburned.

The data in Figure 10-5 falls into two broad bands, with the top six fuels producing lower levels of HC than the bottom five. While the differences are not as large as for the CO, above, the best performing fuels produce less than 70 per cent unburned HCs than the worst. The three worst performing tests, all used the same 95RON (85MON) fuel. This included the test with 20 per cent

HC average – 2000-3000rpm

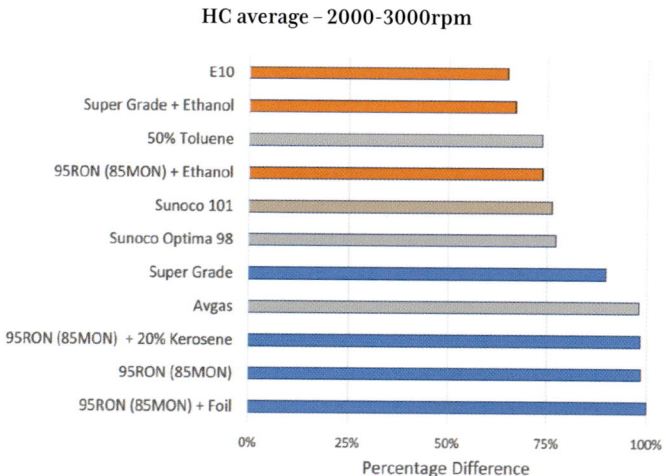

10-5: Unburned hydrocarbons.

added kerosene which shows it burns as well as the unaltered petrol, dismissing any fears that added kerosene will not burn properly and cause damage to the engine.

Relative efficiency

The relative efficiency of the engine was introduced in Chapter 9. It is a measure of the power output divided by the volume of mixture entering the cylinder. The more efficiently the engine runs, the less waste heat it produces and the cooler the engine bay.

Figure 10-6 shows the average full power relative efficiency as a percentage of the best performing fuel. There is a significant difference of 13 per cent between the best and worst. This corresponds to approximately four mpg (miles per gallon) or an extra 1.5 litres per 100km for an engine normally returning 30mpg (9.4 litres per 100 kilometres).

Average efficiency – 2000-3000rpm

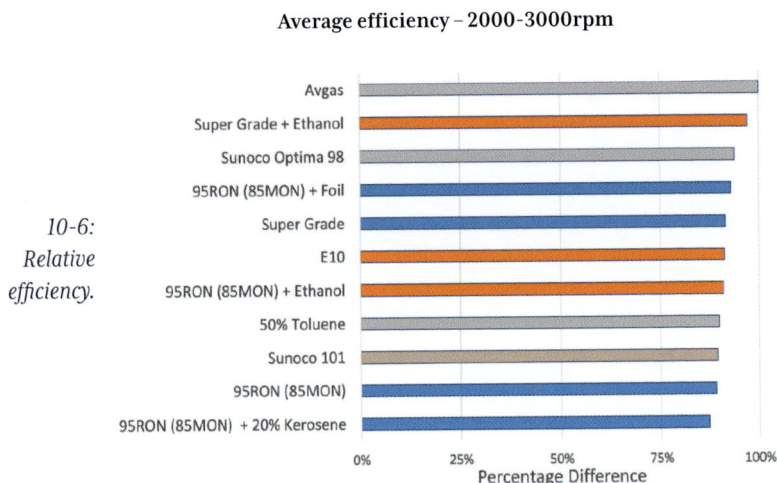

10-6: Relative efficiency.

The reduction in efficiency is due to a high degree of cyclic variability. This data demonstrates how different brands or grades of modern petrol can reduce this effect.

Which is the best fuel?

The test results show there are significant differences in performance between brands and grades of petrol. It is likely other carburettor or classic engines will perform in a similar way. This raises the question, which is the best fuel to use?

It is difficult to know what type of fuel is being sold at any filling station. This makes it impossible to give specific recommendations such as use Brand X, type Y fuel. The fuels can be grouped by factors such as octane rating, ethanol content, etc. The recommendations are based on these groups.

Figure 10-7 shows the overall ranking of the fuels based on the four tests above. For each test, the best performing fuel was awarded 10 points and the worst 0 points.

Ranking – points out of 40

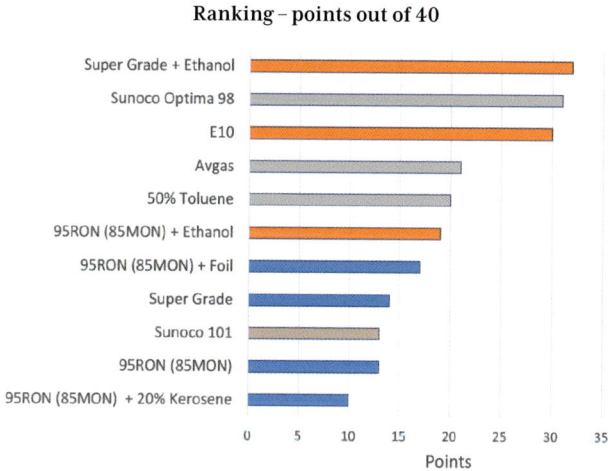

10-7 :Overall ranking.

Two factors stand out. Firstly, octane rating appears to have little effect. Compare for example the Sunoco 101 101RON (90MON) which does not rank as highly as the 95RON (85MON) with ethanol.

This is not surprising. The XPAG is a low compression engine and does not need high octane fuel. Octane rating does not indicate how fast or how well a fuel burns. It only gives an indication of its resistance to pre-ignition.

Two of the three top-rated fuels contained ethanol. Of the top six rated fuels, the three non-specialist fuels contained ethanol. This is discussed further below.

The top rated super grade + ethanol fuel was not tested in the laboratory to determine its volatility profile. The manufacturer's specification sheet suggest the 'summer' version of this fuel will reduce the severity of the Hot Restart Problem. This, or a similar rated product, is probably the best choice for normal use. This ranking supports the claims for Cleveland Discol, the 1930s petrol that contained ethanol, are right.

Warning: Should you plan to use a fuel containing ethanol, read Chapter 3.

Sunoco Optima 98 rated second in both the running tests and in addressing the Hot Restart Problem. This is an ideal choice for low mileage vehicles. With a low front end volatility, it can be stored without 'going off.' Unfortunately, Sunoco Optima 98 is around twice the price of pump fuel. It must be ordered directly from the supplier (see the section at the end of this chapter).

The two worst performing fuels were based on the 95RON (85MON) petrol, the lowest with 20 per cent added kerosene. Should you plan to add kerosene to your petrol to reduce the low temperature volatility, it is advisable to start with a better ranking petrol such as the super grade + ethanol, and add a maximum of 10 per cent kerosene.

Summary

The rankings in Figure 10-7 highlight the problem experienced by carburettor, spark-ignition engines. Of the top six ranking fuels, three contained ethanol and the other three were specialist fuels.

One factor affecting the way an engine runs is how well the petrol and oxygen in the air are mixed before the combustion cycle starts. Ethanol contains oxygen. This is intimately mixed with the petrol. It does not have to rely on turbulent mixing prior to the sparkplug firing.

The oxygen in the ethanol displaces carbon and hydrogen atoms. This reduces the energy level in the inducted fuel. The high ranking of these fuels suggests this reduction is more than offset by the improved combustion.

This same factor applies to the use of the nebuliser (95RON (85MON) + Foil) compared to the 95RON (85MON) fuel on its own. The former is ranked 30 per cent higher even though the nebuliser restricted the area of the inlet manifold. The improved combustion due to the better atomisation and dispersion of the fuel. This more than offsets the restrictions to the flow it causes.

The recommendations for tuning the carburettor(s) in Chapter 11 are based on these results.

Sunoco Optima 98

Sunoco Optima 98 is a specialist petrol. This cannot be bought at a filling station. If you are UK based, it can be ordered direct from the Anglo American Oil Company via their website (www.aaoil.co.uk) or by telephone (01929 551557). For other areas in Europe and in the Middle East, Sunoco distributors can be found on the website above. If you are based in the US, Sunoco's race fuel website (www.sunocoracefuels.com) has got an excellent reseller section where their details can be found.

Sunoco Optima 98 is over twice the price of pump fuel. Its long shelf life, low volatility below 50°C and excellent combustion properties make it the ideal fuel. It is an option worth considering for use in low mileage vehicles or low consumption engines.

Be aware, in the UK the law limits the volume of petrol that can be stored in a garage, or anywhere within six metres of a dwelling, to 30 litres. Other countries may have similar restrictions.

Kerosene

Kerosene, paraffin or 28-second heating oil is a so-called middle distillate of crude oil. Its boiling point is between 150 and 300°C (300-570°F) which overlaps the back-end components of petrol. The effect of adding kerosene is to dilute the front end components, lowering the fuel's volatility below 50°C. Adding kerosene also significantly reduces CO emissions and the degree of cyclic variability. It slightly reduces the engine's power output. The reasons for this cannot be explained.

Adding kerosene is worth considering. Start with low concentrations and small volumes of petrol in the tank. For example, start with five per cent kerosene (one part kerosene in 20 parts petrol), increase this to ten per cent if it appears to improve matters.

Warning: Owners of higher compression engines should take care. Kerosene also reduces octane rating and may cause pinking.

Normally, adding unlicensed hydrocarbons to the fuel of a motor vehicle is illegal. If you live in the UK, you can legally add kerosene to petrol for cars produced before 1956. You will need to apply to HM Customs and Excise for a concession. Write to: Mr John Loughney, Excise, Stamps and Money Businesses, HM Revenue & Customs, 3rd Floor West, Ralli Quays, 3 Stanley Street, Salford, M60 9LA requesting a 'General Licence to mix hydrocarbon oils under Regulation 43 of the Hydrocarbon Oil Regulations 1973 (SI 1973/1311)' and giving your name, address, model and dates of production of your vehicle.

11 Tuning carburettors

The data from Manchester suggests, for road use, that the mixing of the air and fuel before the sparkplug fires is critical in improving the way the engine runs on modern fuel. This chapter discusses how variable jet carburettors can be tuned to improve the way the petrol and air mix.

Warning: Classic vehicles are not all the same. Differences will have accumulated with the passage of time. Also, the severity of problems varies immensely, even between the same models. A solution that may work well for one vehicle may not resolve the same problems in another. The suggestions should be taken as just that; they are not intended as solutions to be blindly adopted.

Turbulence

Perceived wisdom about tuning engines may make an engine worse when driving on the public highway. The main factor influencing the dispersion of the petrol in the inducted air stream is turbulence. This occurs in the carburettor, inlet manifold cylinder head and cylinder itself. The more turbulence, the better the mixing. Unfortunately, turbulent flow reduces the volume of air and petrol entering the cylinder. The less petrol, the lower the power output.

When race tuning engines, the aim is to get the largest possible volume of petrol into the cylinder. This involves gas flowing the cylinder head, matching the inlet manifolds, etc. This reduces the turbulence in the mixture. It is able to flow more freely into the cylinder. The tests showed that above 3000rpm on full throttle, there was already sufficient turbulence to mix the air and petrol. In this case, race tuning the engine will certainly improve performance. Unfortunately, at lower throttle settings and less than 3000rpm it will not.

The test with the nebuliser, discussed more fully in Chapter 8, proves this point. Figure 11-1 shows the difference in power output when the nebulisers were fitted into the inlet manifold. The low reading at 3250rpm is probably caused by the Hot Restart Problem affecting that test run.

Below 3000rpm the nebuliser *improved* the power output of the engine by around one per cent. This is despite it increasing turbulence and restricting the volume of air/fuel mixture flowing into the engine. Something contrary to perceived wisdom.

Above 3000rpm the restriction to the air/fuel mixture flow produces the expected reduction in power.

The steps taken to race-tune an engine may be counter-productive for vehicles primarily intended to be driven on the public highway. A mildly race-tuned engine may run less efficiently under these conditions than an untuned one would.

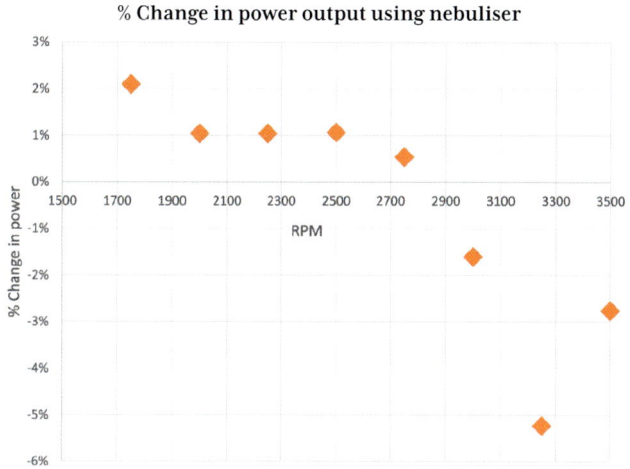

% Change in power output using nebuliser

11-1: Power output comparison with the nebuliser.

While it is not practical to alter the inlet manifold or cylinder head design to increase turbulence. It is possible to tune the carburettors to improve the atomisation and mixing of the fuel. The remaining part of this chapter discusses how this can be achieved. The specific details only apply to the 1 ¼in HS2 SU carburettors, but the general principles also apply to other variable jet carburettors.

This chapter can be read in association with the appendix. This gives a step-by-step guide to rebuilding and tuning twin SU carburettors.

Tuning the carburettor

There are two adjustments that alter the atomisation and dispersion of the petrol in a variable jet carburettor. While both affect the mixture, the change is minor. It can be offset by either adjusting the jet height in the SU or needle height in the Stromberg (see Chapter 5).

The first adjustment is to the height of the petrol in the jet. The second is the speed of the air flowing through the choke.

Petrol height in the jet

Figure 11-2 shows why the height of the petrol in the jet affects the dispersion of the petrol droplets in the choke.

A lower petrol level in the jet has a negative effect on fuel atomisation and dispersion for two reasons.

Low level of petrol in jet   **High level of petrol in jet**

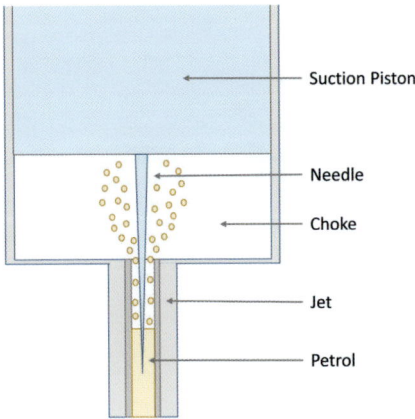

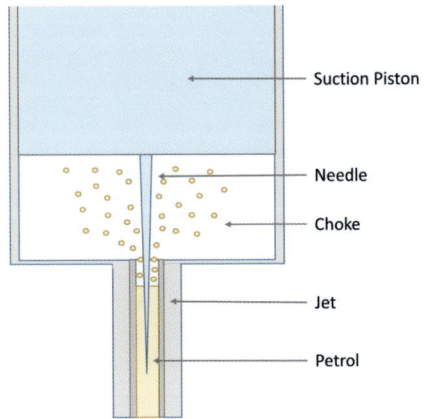

11-2: Effect of petrol height in the jet.

The pressure difference between atmospheric and that in the choke forces the petrol out of the jet. Before it can leave, the petrol has to be raised to the top of the jet. The further the petrol has to be raised, the less force there is to push it out of the jet and break it into droplets.

Secondly, the nearer the petrol is to the top of the jet, the wider the spray as it leaves, better dispersing it into the choke.

There is a balance. If the petrol level in the jet is too high, it will overtop the jet and flood the carburettor.

The petrol height is controlled by the weight of the float and by the setting of the forks in the float chamber. Normally this is adjusted by inserting a rod between the lid and the inside curve of the hinged lever.

When the lever is bent to the correct setting, the needle valve should just close as the forks meet the rod.

With the HS type float chamber use a $\frac{5}{16}$in (8mm) rod (shown in Figure 11-3), and with a hinged nylon float use a $\frac{1}{8}$in (3mm) rod.

However, there is a problem. This time not caused by modern petrol.

11-3: Setting the forks in the float chamber.

It is not always possible to set the correct petrol height in the jet using this method.

The factory handbook for the MG TC recommends a fuel level between ⅛in (3mm) and ⁵⁄₁₆in (8mm) below the jet bridge. The reason this cannot be achieved is because the float is too light. With an HS type float chamber and the correct fork setting, a float weighing 28-30g (1-1.1oz) is needed to achieve the recommended fuel level.

Modern brass floats can weigh as little as 22g (0.78oz). Plastic stay-up floats 20g (0.7oz). The original needle valves were solid steel bars, forks were made out of 20 gauge steel. As the needle valve and forks sit on top of the float, they also add to its weight. Modern needle valves are lighter, made from nylon, aluminium or brass. Modern forks are made from thinner steel. All this adds up to a reduction in the effective weight of the float. Unless any adjustments are made to compensate for these lighter floats, they will lower the petrol level in the jet.

Note: All floats made by Burlen Fuel Systems Ltd are to the original drawing specification of 20-24g (0.7-0.8oz).

The reason a lighter float causes a problem is not immediately obvious. When a float is placed in petrol, it will sink to the point where the mass of the displaced petrol is equal to the mass of the float. The heavier the float, the lower it sinks. The distance between the level of the petrol and the top of the float depends on its weight. The forks sit on top of the float, the lighter the float is, the higher it sits in the petrol and the greater the distance between the forks and the petrol. This lowers the level of the petrol in the float chamber, and consequently in the jet.

No specific tests were run at Manchester to assess the effect of petrol height in the jet. The first set of tests run by the students used the standard weight floats and fork settings. These would have given a petrol height about ½in (13mm) below the jet bridge. The second set of tests were run with standard fork settings but using heavier floats to give a fuel height of ³⁄₁₆in (about 5mm) below the jet bridge.

The later tests showed an average increase in power output of five per cent for three different fuels. This figure should be taken with caution. The tests were run many months apart and other factors may have influenced the measurements.

Figure 11-4 was taken during the first set of tests. It shows the inlet to the carburettor, the bottom of the suction piston and needle. The artificially coloured red petrol can be seen leaving the jet as a stream, rather than a dispersed mist, something not seen in the second set of tests.

11-4: Petrol leaving the jet.

It is possible to compensate for the reduction in piston weight by bending the forks. Bend them towards the lid to raise the fuel level, away to lower it. However, care must be taken not to bend them up too far or the float will foul the chamber lid and cause flooding.

After adjusting the forks, insert a pencil through the hole in the lid and the centre of the float. Check the float does not foul the pivot support before the needle valve shuts. This is highlighted in Figure 11-5.

11-5: Float fouling the lid.

11-6: Washer on float chamber lid.

With the HS2 or other semi-downdraft carburettors, the position of the float chamber relative to the carburettor body changes the fuel height in the jet. Moving it towards the engine lowers the fuel level. Moving it away from the engine raises it.

Ensure the float chamber is positioned with the connecting arm at right angles to the carburettor body.

The suggested way to achieve the recommended fuel level is to increase the weight of the float.

This can be done by carefully adding solder to the base of the brass floats. Stay-up floats are solid plastic and their weight can be increased using self-tapping screws in the base. In both cases, digital kitchen scales are sufficiently accurate to measure the weight.

It is important the float chambers are open to atmospheric. It is easy to think the purpose of the pipes fitted to the top of the float chambers is to remove any petrol that overflows. This is not true, they also act as breather pipes allowing the air pressure in the float chamber to remain at atmospheric.

Check these pipes are not blocked and the correct stepped washer is fitted between the lid of the float chamber and the boss on the pipe. This is shown in Figure 11-6.

Suction pistons

Ensuring the petrol height in the jet is correct will not have a negative impact on performance. This next section is counter to normal practice when race tuning engines.

Early SU carburettors had heavy brass or bronze suction pistons. Later carburettors had aluminium suction pistons with a steel insert. For the 1¼in SU carburettor these weigh 240g (8.5oz). From around 1950, 110gm (4oz) light aluminium pistons were fitted to the 1¼in SU carburettor with a long spring to increase their weight. The red spring increases the effective weight of the suction piston by 128g (4.5oz). The light blue spring by 71g (2.5oz). If it has not worn off, springs can be identified by a coloured band on one end.

Stromberg carburettors are also fitted with a light suction piston and spring.

The advantage of a lighter suction piston is that for a given throttle setting and engine rpm, it will float higher than would a heavier piston. This increases the area of the choke, reducing the restriction to the air flowing through the carburettor. It is one modification favoured when race tuning engines. The disadvantage is that it reduces both the pressure difference across the jet and the velocity of the air flowing through the choke. This in turn reduces the degree of fuel atomisation and dispersion.

Either a fixed weight piston or a light piston with the red spring fitted is suggested for road use.

It is understood that some re-manufactured carburettors are fitted with light blue springs. A suggestion is to remove the suction chambers and check which spring is fitted. If you do, be very careful not to bend the needles!

Effect of piston weight on mixture

When researching the effect of different springs on mixture, I found very little information on the internet; the purpose of this section is to clarify this point. *Note: It is worth having read Chapter 5 before continuing.*

Suction piston weight affects the mixture in two opposing ways:

The lighter the suction piston, the higher it floats for a given volume of air passing through the carburettor. This both reduces the choking effect of the piston and *increases* the size of the annulus between the needle and jet. Making the mixture *richer*.

In contrast the pressure difference between the choke and atmospheric is reduced. This *decreases* the force that is pushing the petrol out of the jet, making the mixture *weaker*.

Figure 11-7 shows the relative effects on the mixture between the fixed weight piston, a lighter piston with the blue spring and a lighter piston with the red

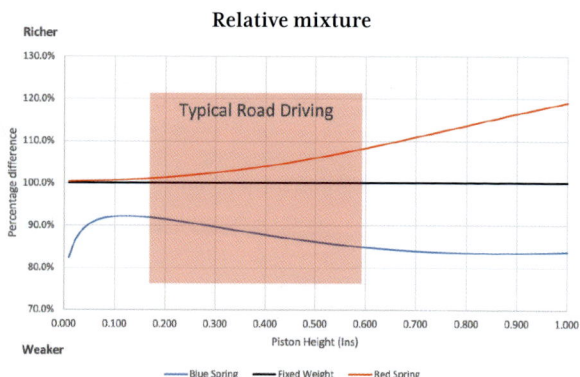

11-7: *Effect of suction piston weight on mixture.*

spring. The data is for the 1¼in SU carburettor using an ES needle.

Both the blue and red springs change the mixture profile with increasing revs/load in the same way as changing the needle would. Fitting a blue spring also makes the baseline mixture weaker. In this case, the jet will need to be adjusted to make the mixture richer to compensate.

For the MG T Types and for normal driving, the ES needle can be used in all three cases. Carburettors fitted with a blue spring can be made richer by screwing the jet-adjusting down. This 'moves' the whole blue curve upwards. Little adjustment is needed for those fitted with red springs.

One problem with the blue spring is that when using the ES needle, the mixture becomes weaker as the engine revs and load increase. A more conservative approach is for the mixture to become richer. The extra unburned petrol keeps the exhaust valves cooler, helping to prevent damage. Using a red spring produces a more conservative mixture curve than with the blue spring.

Summary

The tests identified the engine ran differently above and below 3000rpm. This difference was primarily due to poor mixing and dispersion of the petrol in the air.

Unfortunately, the steps taken to race-tune an engine may make matters worse below 3000rpm. These will reduce turbulence in the inlet manifold, the main factor influencing the dispersion of the petrol in the inducted air stream.

This brings into question what steps should be taken to improve an engine that is primarily used on the public highway. This chapter has suggested ways variable jet, and in particular SU, carburettors can be adjusted to help mitigate this problem.

12 Tuning the ignition system

The Slow Combustion Problem was introduced in Chapter 8. It is caused by a high degree of cyclic variability, particularly at less than 3000rpm. Cyclic variability smears out the time when the peak combustion pressure occurs (Chapter 5), resulting in some cycles burning late. Late-combusting cycles raise the temperature of the exhaust gases which in turn increase the temperature of the cylinder head, exhaust manifold and ultimately the engine bay.

This chapter discusses how the ignition timing can be adjusted to mitigate the effects of this problem. While the example data presented in this chapter applies to the 7.25:1 compression ratio XPAG engine, the suggestions are widely applicable. **Warning: Classic vehicles are not all the same.** Differences will have accumulated with the passage of time. Also, the severity of problems varies immensely, even between the same models. A solution that may work well for one vehicle may not resolve the same problems in another. The suggestions should be taken as just that; they are not intended as solutions to be blindly adopted.

Running an engine with the incorrect ignition timing **can cause damage**. If in doubt, get your car checked by a professional. You can also check https://classicenginesmodernfuel.org.uk/bestfuel/ where owners can rank the brands and grades of fuel that best suit their vehicles.

Centrifugal advance curve

On the time scale of a running engine, after the sparkplug has fired, it takes a long time for the mixture to burn and reach peak pressure. To allow sufficient time for this to happen, the sparkplug needs to be fired before the piston reaches the top of its stroke. This is called 'ignition advance.' Simplistically, the time to reach peak pressure is constant and independent of rpm. Hence, as rpm increases, and the piston is moving faster, the sparkplug needs to be fired earlier in the cycle to give the same time for the fuel to burn. A graph showing the ignition advance against rpm is called the 'advance curve' or 'centrifugal advance curve.'

In practice, as engine rpm increases, so does the turbulence in the cylinder, and the less time it takes for the petrol to burn. Advance curves generally increase as a straight line below 2000rpm, flattening to a constant value as rpm increases. While the general shape of the advance curve is the same for all engines, their detailed shape can vary.

Vacuum advance

The centrifugal advance curve sets the degree of ignition advance based on rpm. The speed at which the flame front grows is also dependent on the pressure of the mixture in the cylinder when the sparkplug fires. Driving on part throttle controls this pressure rather than compression ratio.

The less the throttle is opened, the lower the pressure and the longer it takes for the flame front to grow. More ignition advance is needed to ensure all the mixture has combusted by 17° after TDC. This is achieved by further advancing the ignition timing when the cylinder pressure is low. This is called the vacuum advance.

Importance of the correct ignition advance

It is important an engine is running with the correct ignition advance at any given rpm and cylinder pressure. Either an over-advanced or over retarded ignition damages the engine.

If the timing is too retarded (ie the spark occurs too late), there is insufficient time for the petrol to burn and overly hot gases leave the cylinder. These can burn the piston crown, exhaust valves and damage the cylinder head.

If it is too advanced (ie the spark occurs too early), the peak cylinder pressure will occur when the piston is near the top of its stroke. This places a high load on the big end bearings. It can also result in pinking or knocking.

Pinking is a mechanical tinkling sound that occurs typically at full throttle and low rpm. It sounds like pebbles being shaken about in an empty tin. It is the result of multiple ignition points in the cylinder, rather than the one created by the sparkplug. The noise is from these separate flame fronts 'colliding.' There are two possible causes:

- The first, and the most probable, is because the ignition timing is set too advanced. As the mixture burns, the pressure in the cylinder rapidly increases. A high pressure can spontaneously create secondary combustion points.
- The second cause is glowing carbon deposits in the cylinder, heated by the previous cycle. These may be sufficiently hot to ignite the mixture before the sparkplug fires. This can be made worse by retarded ignition or the Slow Combustion Problem.

Pinking is not necessarily due to an over advanced ignition; it may be caused by the ignition timing being too retarded. It may also be a sign the engine needs de-coking.

The octane rating of the petrol is a measure of how susceptible it is to spontaneous ignition. The higher the octane rating the less likely the creation of secondary combustion points. However, once ignited, higher octane petrol burns at the same rate as one with a lower octane.

How is the ignition advanced?
Centrifugal advance

In modern vehicles, the engine management system controls the ignition timing electronically. In older engines, this is done by a mechanical distributor.

In a mechanical distributor the ignition advance is achieved in two ways. Firstly by what is called the static advance, ie the degree of advance when the engine is not running. Secondly, by bob weights situated underneath the baseplate.

As rpm increases these weights fly outwards rotating the cam that opens the points. These are shown in Figure 12-1. The weaker spring at the top of the photograph should always be in tension. It allows the weights to fly out quickly at low rpm.

When the thicker spring with the looped ends (at the bottom) engages, the rate of advance is slowed. Finally when the weights hit their stops around 3000, there is no further advance. This creates a three-step

12-1: Bob weights in a mechanical distributor.

centrifugal advance curve. It is important this mechanism is working properly, otherwise the engine will be running with the incorrect advance.

The total advance at any rpm is the addition of the static and centrifugal advance timings.

Measuring the centrifugal advance curve at home is relatively easy. All you need is an assistant and a timing light with an advance facility such as those found by searching online for 'advance timing light with advance control.' If your car is not fitted with a tachometer, you will need a more sophisticated timing light that also gives an rpm reading. Ear defenders are also recommended.

Early MGs have a notch in the crankshaft pulley and a pointer on the engine cover to show top dead centre (TDC). Different models of vehicles have similar arrangements. Clean both of these and add a dab of white paint. Disconnect the vacuum advance (if there is one fitted) and block the pipe. Start the engine and ask the assistant to set the engine to run at 1000rpm. When the engine is running steadily, adjust the advance setting on the timing light so the mark on the pulley and the pointer on the engine coincide. You can then read off the ignition advance for that rpm from the timing light.

Warning: Be careful to avoid any moving parts such as the fan, fan belt, or dynamo pulley.

Repeat at 500rpm intervals up to 3500rpm or 4000rpm. You can register and login in to https://classicenginesmodernfuel.org.uk/ignitionadvance/ where you can enter your data, plot your advance curve, and compare it with those from similar vehicles.

This is a very useful exercise. It confirms your centrifugal advance is working. By comparing your advance curve with other published curves, you can validate if the springs in your distributor are the correct ones.

Adjusting the centrifugal advance curve

There are two ways the ignition advance curve can be adjusted.

The first and simplest way is to change the static advance by bodily rotating the distributor. This either advances or retards the centrifugal advance curve by the same amount over the whole rpm range. On some vehicles there is a vernier underneath the distributor. This allows small adjustments to be made. On other vehicles, it is necessary to undo the distributor clamp bolt and CAREFULLY rotate the body by a small amount.

Rotating in the direction of rotation (anti-clockwise on the XPAG) retards the ignition. Conversely rotating against the direction of rotation (clockwise) advances the ignition.

If you adjust your timing in this way, always use the timing light to recheck it. For example at 1000rpm, depending on the type of engine, the advance should be approximately 5° to 10°.

To alter the shape of the advance curve it is necessary to change the weights or the springs. Should the shape of the advance curve need to be changed, it is advisable to get this work done by a specialist.

With programmable electronic distributors, the advance curve can be set using a computer.

Vacuum advance

The majority of vehicles built after the mid-1950s were fitted with a vacuum advance as standard. This consists of a pod on the distributor connected by a fine tube to the inlet manifold or carburettor(s) (Figure 12-2). At low throttle settings, the pressure in the inlet manifold is below atmospheric. This causes the vacuum pod to advance the timing by rotating the plate on which the points are mounted.

Usually, the pods are marked with three digits, eg 5-13-10. This indicates that vacuum advance starts at five inches of mercury (inHg/0.17Bar), ends at 13inHg (0.43Bar), and produces 10° of distributor advance. As the distributor rotates at half the speed of the engine, this corresponds to 20° engine advance.

With a mechanical distributor, the only way to alter the vacuum advance is by changing the pod. On programmable electronic distributors this is done on the computer.

As with the centrifugal advance curve, it is important to check the vacuum advance is working properly. A simple test is to re-measure the advance curve, as described above, this time with the vacuum advance connected. As the measurements are taken with the engine running on a light throttle, this curve should be 5°-15° in advance of the centrifugal one.

12-2: Vacuum advance pod.

Superchargers

A warning for those cars fitted with superchargers: Superchargers INCREASE the pressure of the gases in the inlet manifold above atmospheric. This causes the flame front to grow more quickly requiring LESS advance. Setting the vacuum advance on these engines is more difficult. The pod on a mechanical distributor may not cope with positive manifold pressures.

Enhanced ignition systems

There are products on the market that increase the power of the spark and claim to improve the engine's power or efficiency. This was investigated at Manchester. The full throttle power output was measured as the voltage applied to the coil was changed. The lower the voltage on the coil, the less energy there is in the spark.

Figure 12-3 shows the results of these tests. They demonstrate that spark energy has no effect on power output over the range of rpm investigated.

The chemical energy released by the small volume of petrol ignited by the spark is over 1000 times that of the electrical energy in the spark itself. As long as a spark is formed, no matter how weak, the combustion process will be triggered.

The only practical benefit of systems to improve the spark energy is to reduce misfiring. Misfiring or missing is usually the result of defects in the high tension leads or sparkplugs. Even high energy systems will fail to deliver a spark to the plug if the condition of the ignition system is poor. The distributor cap, rotor arm, coil, plug caps and plug leads should be

Effect of spark energy on power output

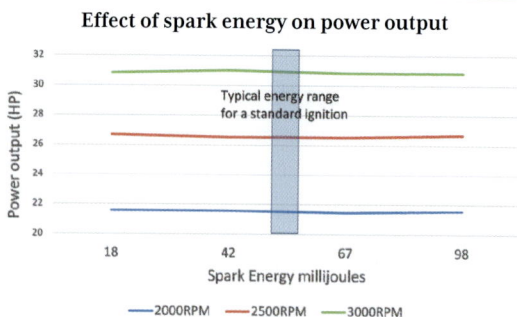

12-3: Effect of increased spark energy.

regularly inspected. They should be mechanically sound (no scratches or burns from arcing) and free from dirt and moisture.

One problem with modern petrol is that it can leave hard carbon deposits on the sparkplugs. These can build up and short the spark. The sparkplugs should be checked and cleaned if necessary. The plug gap should also be checked. This is usually set to 0.025in (0.6mm). As a quick test, this is the average thickness of a thumb nail.

Regular maintenance will ensure even a standard mechanical ignition system will perform optimally.

Electronic ignition systems claim to offer significant improvements. These include no points to wear, a stronger spark and improved timing accuracy.

It is probably not worth fitting an electronic ignition system. They offer little practical benefits. The owners of classic vehicles often cover only a few thousand miles every year. At this rate, a mechanical distributor fitted with a good quality condenser will run for a number of years without needing to be maintained. The tests above show spark energy does not improve engine performance. Finally, cyclic variability adds around 5° to 10° error to the timing accuracy. This is larger than the 1°-2° timing error of a mechanical distributor.

The disadvantage of electronic ignition systems. A more complex system that cannot be fixed at the side of the road should it fail.

Manchester advance curves

One aim of the Manchester tests was to measure the centrifugal advance curve for the XPAG engine. Comparing this with the original curve makes it possible to identify any differences caused by modern petrol.

While these measurements only apply to the XPAG, experience has shown the suggested changes also apply to other engines.

For every test, the timing was set to give the maximum power. From this data it is possible to calculate the ideal centrifugal and the vacuum advance curves.

Figure 12-4 compares the centrifugal advance curve from the Manchester data (blue curve) with the original (red curve). For the original curve, the static advance is set to that recommended in the manufacturer's handbook.

This reflects the previous findings. Below 3000rpm the engine does not run as well as it did originally. Up to 3000rpm, the original advance curve is too retarded. Above 3000rpm it becomes a better match to that measured at Manchester.

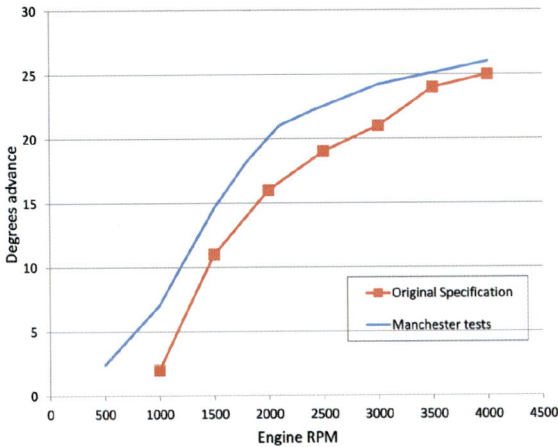

12-4: Measured v/s original centrifugal advance curve.

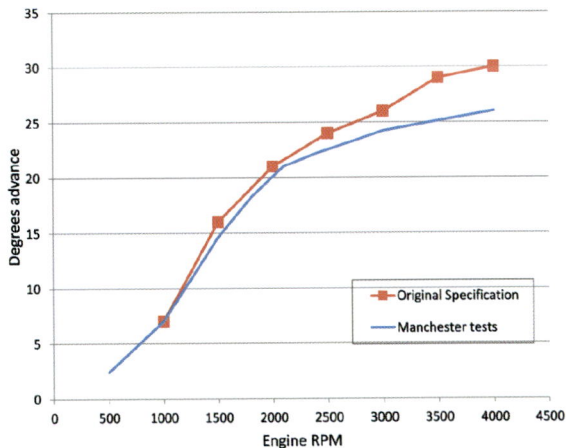

12-5: Original advance by 5°.

Figure 12-5 re-plots the data shown in Figure 12-4 with an additional 5° static advance on the original curve. This is achieved by bodily rotating the distributor against its direction of rotation.

While this corrects the problems below 3000rpm, it over-advances the engine above 3000rpm. At these engine speeds the additional four to five degrees of advance is less serious than running retarded at lower rpm. High engine rpm is used less frequently than low rpm during normal driving.

The need for the 5° advance up to 3000rpm provides supporting evidence of the high degree of cyclic variability. This was discussed in Chapter 6.

In common with many prewar engines, the XPAG was not fitted with a vacuum advance. However, the vacuum advance curve was measured at Manchester and is shown in Figure 12-6.

These show an additional 5°-15° of advance can be added to the centrifugal advance on low to medium throttle settings, normally used when driving on the public highway.

Vacuum advance curve

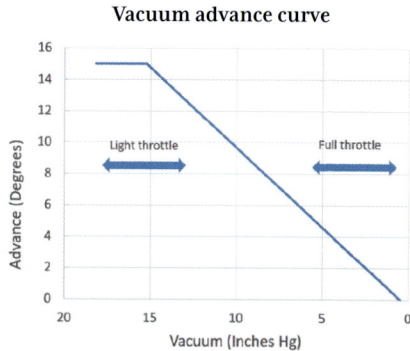

12-6: Vacuum advance curve.

Exhaust gas temperature and ignition advance

When an engine experiences a high degree of cyclic variability, advancing the ignition timing has a significant effect on exhaust gas temperature.

This is demonstrated in Figure 12-7, which plots two peak pressure frequency curves. The blue curve shows a normally advanced ignition. The orange curve an overly advanced ignition. (See Chapter 6 for a description of the peak pressure frequency curve). Advancing the ignition timing moves the curve to the left.

The change in the number of cycles occurring in the maximum power band is small, as is the increase in early-combusting cycles (red area). Advancing the ignition timing by a small amount does not alter the power output or significantly increase the risk of pinking.

In contrast, there is a large reduction in the number of late-combusting cycles (green area).

Figure 12-8 shows the measured change in power output as a function of ignition advance. As predicted, once peak power is reached, it falls slowly as

Effect of advancing ignition timing

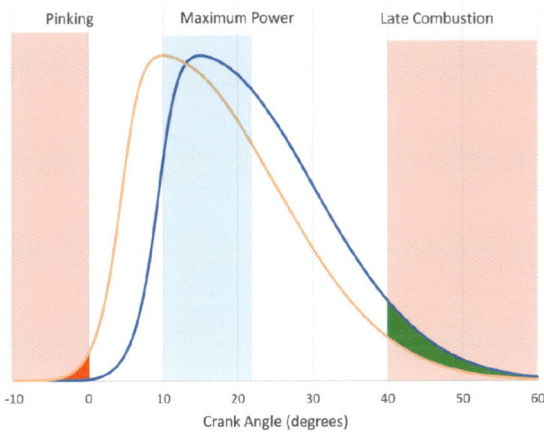

12-7: *Effect of advancing ignition on exhaust gas temperature.*

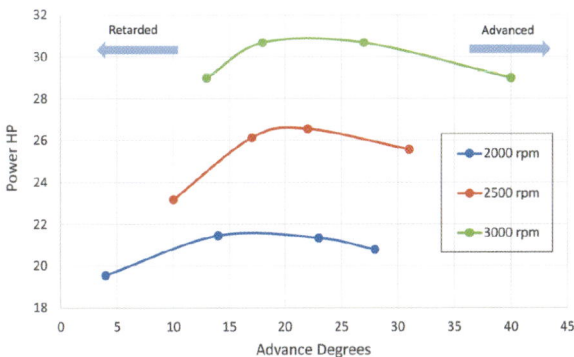

12-8: *Change in power output with ignition advance.*

the ignition timing is advanced. (Right-hand side of the graph).

The same is not true of exhaust gas temperature. This is shown in Figure 12-9.

As ignition timing is advanced, the exhaust gas temperature falls significantly. This is because of the large reduction in late-combusting cycles. The average reduction in temperature is 13°C (55°F) for each 5° advance.

Using the manufacturer recommended advance setting, which is 5° too retarded, increases exhaust temperatures by 15°C. This is the reason why people believe modern petrol 'burns hotter.' The fact that this myth is widespread suggests this problem is seen in many classic engines.

The hot exhaust gases from the late-combusting cycles are why some owners experience burned pistons and exhaust valves or cracked cylinder heads.

When setting the ignition timing it is better to err on the side of being too advanced than too retarded. This will help reduce exhaust gas and engine bay temperatures with only a minor loss of power. But beware of pinking.

The vacuum advance allows the ignition timing to be further advanced without the risk of pinking or knocking. Fitting a vacuum advance to those engines without one is something that is recommended. Owners of early MGs who have fitted one have reported significant improvements to the way the engine runs.

Exhaust temperature and function of ignition advance

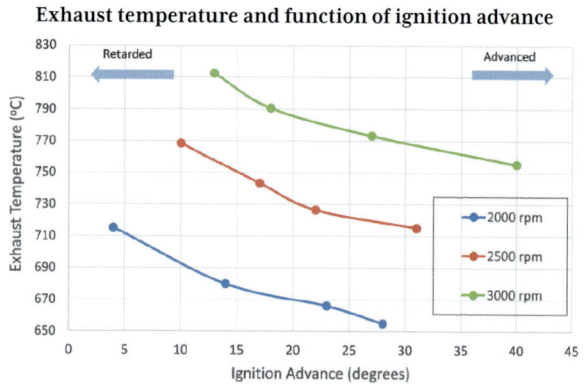

12-9: Temperature v/s advance.

Summary

Ensuring the engine runs with the correct ignition advance over the whole usable rpm range is very important. If it is too advanced or too retarded the engine may be damaged. Both the centrifugal advance curve and vacuum advance curve (if fitted) should be checked and compared to those recommended for that engine. The operation of the bob weights in the distributor should be checked as part of the regular service. If a vacuum advance is not fitted, it is worth considering.

Chapter 13 discusses fitting a vacuum advance. The first section describing the options for connecting the vacuum pipe is generally applicable.

There was little evidence that high energy or electronic ignition systems delivered any practical benefits.

The tests showed the shape of the manufacturer-recommended centrifugal advance curve was appropriate for modern petrol. However, it was 5° too retarded below 3000rpm. It is probable other engines will be affected in this way, but not necessarily to the same degree.

The data also showed that power output is not very sensitive to ignition advance. In contrast, the exhaust temperature is significantly reduced as the ignition timing is advanced. Providing the engine is not pinking, advancing ignition timing will reduce exhaust and hence engine bay temperatures. It will also reduce possible engine damage.

13 Fitting a vacuum advance

Chapter 12 discussed the importance of having the correct ignition advance. It suggested it was beneficial to advance the ignition timing as much as possible to reduce exhaust gas temperatures.

Chapter 12 also discussed the vacuum advance. This further advances the ignition timing on low throttle settings without the risk of pinking. A vacuum advance was fitted as standard to most vehicles built after 1960.

Classic carburetted engines do not run properly below 3000rpm on light throttle when using modern petrol, driving typical of normal road use. This is when a vacuum advance will operate. Fitting a vacuum advance to engines without one is something worth considering.

Feedback from those who have fitted vacuum advances to their MG T-Types has been positive. Engine temperatures are noticeably lower in slow moving and stop-start traffic. The engine runs cooler when cruising on the flat up to around 35-40mph. Overall response is better.

This chapter discusses the available options when fitting a vacuum advance. While the examples are for the XPAG engine, the majority of this chapter is relevant for any engine that does not have a vacuum advance.

Vacuum take-off

The pressure in the inlet manifold, on the engine side of the throttle butterfly, depends on throttle setting. At full throttle the pressure is virtually the same as atmospheric pressure, dropping to a near vacuum as the throttle is closed. The full throttle pressure will exceed atmospheric for engines fitted with superchargers. The pressure difference between atmospheric and that in the inlet manifold is what drives the vacuum advance.

The vacuum feed must be connected to a take-off point on the engine side of the carburettor butterfly. There are three possible options:
- Use a take-off on the carburettor.
- Adapter plates fitted between the inlet manifold and carburettor.
- The inlet manifold.

Each of these options has different characteristics that need to be considered. Specifically: responsiveness to changes in pressure, pickup from tick over, and susceptibility to back pressure pulses.

Damper

Chapter 8 discussed how cyclic variability can result in pressure pulses in the inlet manifold. A responsive vacuum take-off point will pick up these pressure pulses, possibly giving the wrong vacuum advance.

Even if your car is already fitted with a vacuum advance it is worth investigating if the vacuum take-off is susceptible to pressure pulses. To do this, buy a cheap vacuum gauge (search online for 'vacuum gauge car'). Fit it to the distributor end of the vacuum tube. Run the engine between 1000rpm and 2000rpm. If the needle vibrates this shows that pressure pulses are present. In this case, it is advisable to fit a damper between the take-off point and distributor. These are commercially available and can be found online by searching 'vacuum advance anti pulse.'

Dampers are not complicated. I have made one from a piece of 15mm copper pipe (Figure 13-1). This has two connectors on one end to take the feeds from the twin carburettors. The other end has a single connector to the distributor with a fine 1mm diameter hole.

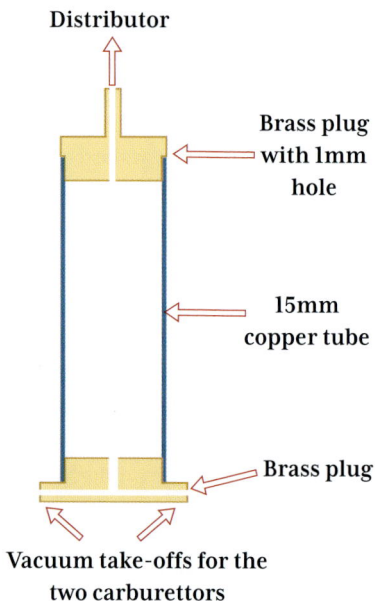

13-1: Damper.

Response from tickover

When the engine is ticking over, the throttle is virtually closed. The pressure in the inlet manifold is very low, giving the greatest ignition advance.

When the throttle is pressed, more air/fuel mixture will enter the cylinder. This needs less ignition advance. If the distributor is not able to respond quickly enough, the engine will run too advanced. This may cause pinking or knocking, uneven running or even stall the engine. To test if this is a problem, when the engine is ticking over, briefly blip the throttle fully open and listen to the pickup. If the engine sounds rough, it may be due to a the distributor responding too slowly.

On some electronic programmable units, it is possible to set an rpm below which there will be no vacuum advance, avoiding this problem completely. Unfortunately, there is nothing that can be done with a mechanical distributor other than changing the vacuum take-off point (see below).

Carburettor body take-off

One possible take-off point is the carburettor body. Later SU carburettor bodies already have a hole drilled into them. These are normally blanked off. Burlen Fuel Systems Ltd sell an adapter that will screw into this hole. The outer position of this fitting is shown in Figure 13-2 and the point at which it enters the carburettor in Figure 13-3.

13-2: Vacuum take-off.

13-3: Vacuum position of hole.

13-4: Vacuum pipe fitting.

Figure 13-4 shows the black vacuum pipe running from the rear carburettor of the XPAG and behind the block to the distributor.

The advantage of using this fitting is that it provides a rapid response to changes in inlet manifold pressure. At tickover, the hole is blocked by the butterfly which prevents over advancing the engine on initial pickup. It may be advisable to fit a damper. This is probably the best take-off point to use with a mechanical distributor.

Classic Engines, Modern Fuel

Adapter plates take-off

At Manchester, specially made adapter
plates (Figure 13-5) were fitted between
the carburettors and inlet manifold.
These were used to measure the
manifold pressures. They were modified
by removing the vacuum gauge and
thermocouple to make two vacuum
take-off points, one on each carburettor.
Similar adapter plates are available from
minispares (part numbers: MFA132,
MFA338, and MFA446 for the different
sizes) (Figure 13-6).

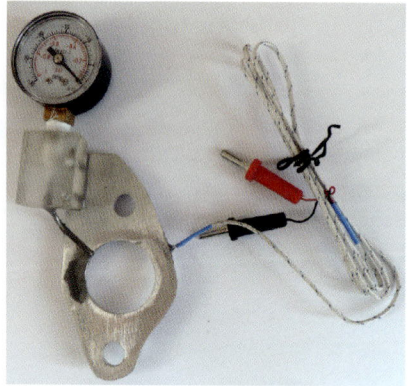

13-5: Adapter plate.

Chapter 9 suggested fitting thermal
spacers between the inlet manifold and
carburettors. One option may be to fit
a vacuum take-off to these by drilling a
fine hole and fitting a connector.

The advantages of adapter plates
is that they provide a rapid response
to changes in pressure. No permanent
alterations are needed. A damper
should be considered. This take-off
is very susceptible to pressure pulses
in the inlet manifold. At Manchester
the vacuum gauges were seen to

13-6: Commercial adapter plate.

be vibrating in some of the tests. Unlike the
carburettor take-off, there is no mechanism to
prevent over advancing the engine from tickover.
This is probably the easiest solution to use with an
electronic distributor or external electronic box.

Inlet manifold take-off

Many later engines took their vacuum feed directly
from the inlet manifold. It is possible to drill, tap
and fit a vacuum take-off directly into the inlet
manifold. This change cannot be reversed. On the
XPAG it is possible to fit the vacuum take-off using
a modified core plug in the balance tube. Core

13-7: Modified core plug.

114

plugs are available from NTG Motor Services Ltd. This is shown in Figures 13-7 and 13-8.

A similar solution may be suitable for other engine types.

An inlet manifold take-off will provide a slower response to rapid pressure changes. This will remove the need for the damper. However, the slow response may cause problems with engine pick-up from tickover.

13-8: Manifold take-off.

Distributor options

There are three suggestions for adding a vacuum advance to the engine.

- Replace the existing mechanical distributor with one that has a vacuum advance pod.
- Replace the existing mechanical distributor with an electronic version that has a vacuum sensor.
- Keep the existing mechanical distributor and fit an external ignition control box.

Mechanical distributor for the XPAG

Options for the XPAG engine are to use an MGB 25D4 or 45D4 or a Metro 59D4 distributor, all of which are relatively easy to source. For the XPAG engines, it is necessary to replace the drive dog with a gear (available from NTG Motor Services Ltd). Figure 13-9 shows the later distributor on the left, alongside the original distributor with the drive gear. The drive dog is held in place by a roll pin. This is easily removed

13-9: Lucas 25D4 distributor with vacuum pod, alongside the original.

allowing the dog to be pulled off the shaft. In the photograph the roll pin on the right-hand distributor is shown partially removed.

These distributors will fit directly into later TDs and TFs (ie those fitted with the D2A4 distributor and without the adjustment vernier). They will need to be machined to fit the earlier XPAG engines that use the DKY4A unit. The distance between the base and the drive gear must be increased by approximately 5mm to accommodate the thickness of the vernier. A groove must also be machined into the shaft to take the locating bolt.

The problem with the XPAG engine is that there is very little physical space for the distributor. It is located between the engine block, breather pipe and rear of the dynamo with its rev-counter drive. Figure 13-10 shows a 45D4 distributor fitted to a 1500 MG TF. This has a relatively small vacuum pod and it fits well in the confined space.

Other engine types may suffer from similar problems.

The advantage of a mechanical distributor is that it is simple, efficient and easy to fix should it go wrong.

Programmable electronic distributor

Chapter 12 suggested few benefits could be found from fitting an electronic distributor. Some electronic ignition systems offer two genuine advantages: a vacuum advance and the ability to programme the centrifugal and vacuum advance curves.

13-10: 45D4 distributor fitted to a 1500 MG TF

13-11: 123TUNE distributor.

The basic models allow for one of a set of pre-programmed curves to be selected. Higher end models allow bespoke advance curves to be programmed using a computer or phone app.

One possible solution is the 123TUNE which can be supplied as a replacement for the MGB distributor. For earlier XPAG engines, it will need to be modified as described above. Conversions are available from 123ignition-conversions.com. This distributor is smaller in size than the original: with no vacuum pod, just a small brass vacuum tube (seen on the left-hand side of Figure 13-11), it easily fits into the XPAG engine.

This distributor is fully programmable, allowing both the centrifugal and vacuum advance curves to be set. This model also allows positive vacuum settings to be entered, enabling it to be programmed for an engine with a supercharger. It will work with both positive and negative earth vehicles.

Figure 13-12 shows the 123TUNE distributor fitted to a 1250 MG TC.

123TUNE distributors are available for a range of models of cars.

CSI also produce an electronic distributor for the XPAG. This version has a vacuum advance pod, similar to a mechanical distributor. This makes it more difficult to fit.

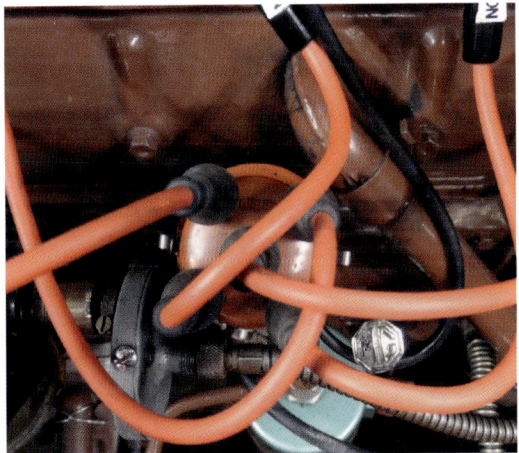
13-12: 123 fitted to a 1250 MG TC.

Programmable black box

An external electronic box can be fitted as an alternative to replace the original distributor. These use the original distributor and points with the centrifugal advance locked. They are triggered by the mechanical points although some units are compatible with electronic sensors. Both the centrifugal and vacuum advance curves can be set using a computer. Possible products are the:

13-13: Amethyst black box.

- Aldon 'Amethyst' Mappable Ignition System (Figure 13-13).
- Interceptor™ Ignition
- CB Performance Black Box Programmable Timing Control Module.

The main benefit is that these units are easy to fit. Should they fail the ignition can be reverted back to using the mechanical distributor. Currently they are only supplied to fit negative earth vehicles. It is possible to make an inverter which will allow it to be used on a vehicle with a positive earth. Inverters are not commercially available.

Summary

A vacuum advance reduces exhaust and engine bay temperatures, especially when driving at slow speeds or in stop-start traffic. Unfortunately, they only became common place for vehicles built from the 1960s.

This chapter discussed the various vacuum take-off points, their advantages and possible issues. It has also suggested some solutions to upgrade engines to provide a vacuum advance.

Where cars already have a vacuum advance, it is important to ensure this works properly.

A point of interest: a vacuum advance is not needed on vehicles that will primarily be used for racing. These will be mostly run on fully open or fully closed throttle. Under these conditions a vacuum advance offers no benefits.

14 Testing an engine's efficiency

Petrol produces a colossal amount of heat energy when it burns. A spark-ignition engine only converts around 30 per cent of this energy into power. The remaining 70 per cent goes into increasing the temperature of the exhaust, engine bay and, for water-cooled engines, the cooling water. The less efficient the engine, the more heat it generates.

Fuels that increase the degree of cyclic variability also reduce the engine's efficiency. It would be ideal if it was possible to measure this effect directly. This would provide a precise means of choosing the best fuel (ie one that reduces the degree of cyclic variability). Unfortunately, this is not practically possible.

This chapter describes a simple, low-cost method that can be used to give an estimate of how much waste energy the engine produces. A single measurement on its own is of little use. By comparing measurements over time with different fuels it is possible to get an indication of which is the best performing fuel.

This may not work with your vehicle, but it is something that is cheap and easy to fit. If it does work, the benefits are huge. It lets you assess the quality of the fuel you are using as you drive along.

Unfortunately, this method is only applicable to water-cooled engines fitted with a thermostat.

Engine thermostats

Thermostats were fitted from the 1940s onwards. They serve two purposes.

Until an engine has warmed up to 90°C-105°C (195°F-220°F) it runs less efficiently, producing more pollution. The faster it can warm up to full working temperature the better.

When an engine is cold, the thermostat is closed. It diverts the cooling water through a bypass rather than through the radiator. This both reduces the volume of water that needs to be heated and prevents the water being cooled by the radiator. When the engine is up to temperature, the thermostat opens, allowing the hot water from the engine to circulate through the radiator.

The second purpose of the thermostat is to keep the engine's temperature constant. It is easy to imagine that once a thermostat has opened, it stays fully open allowing all the cooling water to pass through the radiator. This is not true.

Vehicle radiators are designed to keep the engine cool even when it is running at full power. When driving on the public highway, the engine is running at a part load. It produces considerably less heat than when running under full load. As a result the thermostat only allows enough water into the radiator to maintain the ideal engine temperature.

This can be illustrated by thinking of the engine and radiator as two leaking containers, shown in Figure 14-1. The heat from the petrol is like the tap, 'pouring' heat into the engine (left-hand container). This heat is 'leaking' out through the exhaust system and by radiation from the cylinder block. As the tap is turned up, the rate at which heat enters the engine exceeds the rate at which it is being lost. The engine 'fills up' and its temperature rises.

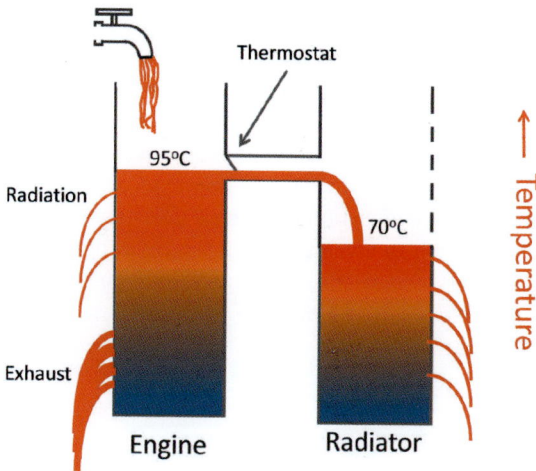

14-1: Engine and radiator.

Once the engine's temperature reaches 95°C (205°F), the thermostat opens, and the heat 'pours' into the radiator. The more heat in the engine, the faster it flows into the radiator. Conversely, the less heat, the slower it flows. When the engine temperature drops below 95°C (205°F) the thermostat closes. No more heat can flow into the radiator.

Unlike the engine, the radiator has a large number of 'heat holes' in it. The hotter it gets the more heat it loses. As the heat from the engine flows into the radiator, its temperature rises to a level where the heat being lost is the same as that coming from the engine. The temperature the radiator reaches needs not be the same as the temperature of the engine. In Figure 14-1 it is shown as 70°C (158°F).

Why is this important? When the engine is delivering the same power output, the less efficiently it does so, the more waste heat is generated. This extra heat flows into the radiator, increasing its temperature. In this way, the temperature of the water in the radiator gives an indication of how efficiently the engine is running.

In practice, matters are not that simple. The heat generated by the engine depends on both its efficiency and how hard it is working. When driving up a hill, for example, the engine will need to work harder and will generate more heat. The temperature of the radiator will rise. When driving down a hill, the engine is working less hard and the temperature of the radiator will fall.

The heat lost from the radiator depends on how much air is flowing through it (the vehicle's speed) and the ambient temperature. On a cold day it will lose heat more quickly than on a hot day. Its temperature will be lower.

With these factors in mind, by monitoring the average radiator temperature over time with a given fuel gives an indication of how efficiently that fuel is combusting. The lower the average temperature, the better the fuel.

Does this work in practice?

The answer is yes. Figure 14-2 shows the radiator temperatures measured using different fuels in my MG TC. These are the averages when driving over a country route and at a steady 60mph (95kph).

All the tests were performed on days with similar ambient temperatures and driving conditions. (Note: these test were run when it was still possible to buy classic 4* petrol from specialist distributors.)

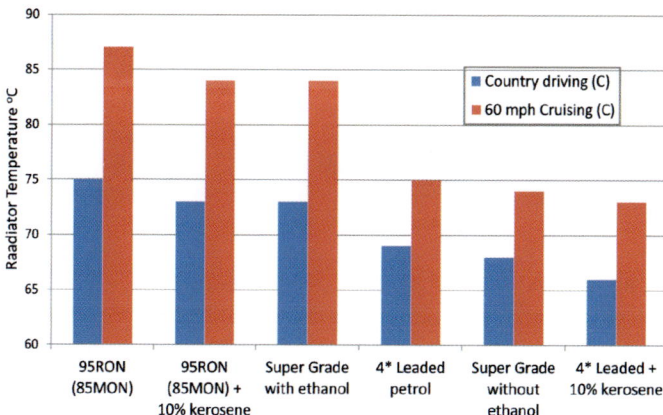

14-2: Radiator temperatures with different fuels.

There is a measurable temperature difference of 10°C (18°F) between the best and worst performing fuels.

These tests were performed some years ago. It was as a direct result of this test that some MG owners changed to super grade fuel. All those who did so reported similar positive results.

How do you measure the radiator temperature?

On the majority of vehicles, the engine temperature gauge is connected to a sender in the engine block. It measures the temperature of the engine (the left-hand container in Figure 14-1) not the radiator temperature.

As standard, the MG TC was not fitted with a temperature gauge. It has a fitting for a sender in the top of the radiator. This is on the radiator side of the thermostat. It measures the temperature of the cooling water in the top of the radiator, not the engine. Figure 14-3 shows the temperature sender for my MG TC.

14-3: Temperature sender on my MG TC.

Few vehicles have fittings for temperature senders in the header tanks of their radiators. A simple alternative is to use a thermocouple placed on to the radiator connected to a temperature gauge inside the vehicle. An inexpensive, electrical multimeter with a thermocouple is ideal (search online for 'multimeter with thermocouple').

Care is needed when positioning the thermocouple on the radiator. The TC has a large header tank and there is enough volume of water to allow the hot water from the engine to mix. The thermocouple should not be positioned where the direct flow of water from the engine enters the radiator. Also ensure it is not in the airflow from the front of the car, otherwise it may give a false reading.

Once installed, the monitoring system is easy to test. With a warm engine, watch the temperature when driving up a hill. After a few minutes it should start to go up, falling again on the way down the hill. After that, watch the gauge over several trips to get some idea of the average running temperature. When you change to a different fuel, keep an eye on the gauge. Hopefully, you will note small differences. Choose the fuel that gives the lowest radiator temperature.

Summary

While there are no guarantees you will get any useful information, fitting a thermocouple is simple and does not cost very much. If it works as well as my temperature gauge, it provides an excellent means to check the performance of different fuels.

15 Conclusion

The aim of this book has been to inform enthusiasts about the complex inner workings of a spark-ignition engine, and why modern petrol causes the problems it does. The findings are mainly based on a comprehensive set of tests run at Manchester University. The suggested solutions to running classic vehicles on modern petrol are summarised in this chapter.

Problems and their causes

The high volatility of modern petrol below 50°C (122°F) is the main cause of what is often referred to as the Hot Restart Problem. At a typical engine bay temperature of 50°C (122°F), only 8 per cent of petrol from the 1960s would have evaporated. In contrast 25 per cent of modern petrol will have evaporated at this temperature.

This high volatility is also the reason modern petrol 'goes off' when stored in a vehicle's petrol tank. The lighter fractions, called front end components, evaporate over time, changing the fuel's composition.

Many classic vehicles suffer from the Hot Restart Problem. When a vehicle is stopped or moving slowly in traffic, the temperature of the engine bay starts to rise. With little or no petrol flowing through the fuel pump, petrol lines and carburettors, it has more time to get hot and boil. Carburettors will not deliver the correct mixture when there are bubbles of vapour in the petrol. This weakens the mixture which causes the engine to stop and prevents it from restarting.

There is a second, less obvious problem. Modern petrol *appears* to burn more slowly and hotter than classic petrol. These symptoms are caused by a high degree of cyclic variability (Chapter 6). This has the same effect as running with a retarded ignition. It increases the temperature of the exhaust gases. These overly hot exhaust gases can damage the engine. They also raise engine bay temperatures and make the Hot Restart Problem worse.

There is a 'sting in the tail.' The tests at Manchester found the degree of cyclic variability is at its worse at engine speeds and loads typical of driving on the public highway.

In practice, a high degree of cyclic variability does not retard the ignition timing. The time it takes for the air/fuel mixture to burn depends on many different factors. Random variations in these factors cause some of the combustion cycles to take longer to burn than the ideal.

Modern petrol appears to make the degree of cyclic variability worse, increasing the number of slow burning cycles. Even with the correct ignition timing, this has the apparent effect of retarding the ignition.

Possible solutions

Unfortunately, there is no single magic fix to the problems caused by modern petrol. There are a number of steps which, when taken together, reduce their severity. These are:

- Use a better performing petrol.
- Reduce the temperatures of the engine bay.
- Stop the heat getting to the fuel system components.
- Tune the engine to reduce the degree and effect of cyclic variability.

Use a better performing petrol

The choice of fuel can have a significant effect on the way an engine runs. This is discussed in Chapter 10. There are two factors:

- Volatility – the more volatile the fuel at typical engine bay temperatures, the more prone it is to the Hot Restart Problem.
- Combustion characteristics – particularly at the engine loads and rpm used when driving on the public highway. Chapter 14 describes a simple means to monitor the way the engine runs on different fuels.

Volatility

The lower a fuel's volatility at typical engine bay temperature, the less likely it will be to cause the Hot Restart Problem.

The only practical way to achieve this is to use a specialist petrol such as Sunoco Optima 98. This is sold by Anglo American Oil. It is expensive but its volatility matches that of 1960s petrol. It can be stored without degrading. It is also ethanol free, and worth considering for low mileage vehicles.

The only other guaranteed way to reduce volatility is by adding kerosene to the petrol, which is legal, in the UK, for cars built before 1956. Kerosene works by diluting the front end components of modern petrol. The more you add, the greater the effect. The down side is that at concentrations above 10 per cent it reduces the power output. It also lowers the petrol's octane rating. If you plan to try this with a high compression engine, you need to take care.

The distillation tests reported in Chapter 2 and 10 must be performed by specialists. Visit https://classicenginesmodernfuel.org.uk/bestfuel/ to see the volatility of the brands, and grades popular with classic vehicle owners.

Should you choose to use standard pump fuel, the volatility changes over the year. It is more volatile in the winter (to make it easier to start the engine) and less volatile in the summer. Try to avoid winter, spring and autumn petrol and only fill up in the summer. Unfortunately, finding a summer petrol is easier said than done. Even petrol bought at the same filling station can vary between deliveries.

Some super grade fuels appear to be less volatile. The addition of ethanol does not appear to make petrol more volatile.

Combustion characteristics and ethanol blended petrol

Of the top six rated fuels tested at Manchester, the three best performing, non-specialist fuels contained ethanol. Ethanol contains oxygen. This improves the mixing of the air/fuel in the cylinder. This, in turn, improves the combustion process, especially at lower rpm. The top performing fuel was a super grade blend that contained ethanol up to 5 per cent.

Ethanol blended petrol is here to stay. Over time, concentrations will rise. While there are issues, it appears that it is not the 'baddie' that some people fear.

There are two practical problems: rotting hoses and seals. More serious is the severe corrosive effects of any water that may get into the petrol. As part of the regular service, owners should check for petrol leaks. Start the car to pressurise the fuel system. Feel around the rubber hoses, carburettors and fuel pump. A dry kitchen towel is a good way to detect leaks.

Be very careful when filling the vehicle with petrol, especially on wet days. Do not get ANY water into your fuel system. Consider slosh coating your petrol tank and possibly the float bowls. Periodically drain the fuel tank and let it dry out. Annually clean out the carburettor float bowls and replace any filters.

Reduce the temperatures of the engine bay

The exhaust system, air heated as it passes through the radiator, and the heat lost from the engine block all raise the engine bay temperature. When driving, the airflow through the front of the car dissipates this heat. It is important to keep the engine bay, particularly the fuel system, temperatures as low as possible (Chapter 9). Check the cooling system to make sure it is working efficiently. Ensure cold air can flow freely around the engine, particularly around the fuel system components.

Flush out the radiator, remove flies and other debris from the radiator fins. Check the thermostat is working. On the older cars it is possible to fit the cooling fan the wrong way around. Consider replacing an old pressed steel fan with a seven-bladed plastic fan like that fitted to an MGB. A wetting agent in the cooling system may also help.

Reposition ancillary equipment such as the horn, badges or additional lamps to ensure they are not blocking the airflow through the front of the car.

Re-route fuel hoses and pipes, especially if they are near the hot exhaust manifold.

Air that has passed through the radiator is hot. In slow moving traffic, an electric radiator fan may make matters worse. This will switch on as the radiator heats up, blowing hot air into the engine bay. It may be better to position it at the bottom of the radiator so it can suck in cooler air.

Fit a timer or similar circuit to keep the fan running for five to ten minutes after the engine has stopped. This will help disperse the hot air from the engine bay. Another possibility would be to add a switch or circuit to reverse the voltage polarity to the fan. This would draw cool air from under the car and vent the hot air through the front of the radiator. It could be used when the vehicle is stopped or moving slowly.

On hot days, think about where you park your vehicle. If parked in direct sunlight, the exposed petrol tanks such as those on motor bikes or the back of older vehicles can get quite hot. Even more modern cars with internal petrol tanks can get very hot in the sun.

Stop the heat getting to the fuel system components

Insulating fuel system components and fitting heat shields may help. These will only slow the transfer of heat, not stop it. Any insulation will need to prevent heat getting to the petrol until the engine bay temperature has had time to drop below 40°C.

The tests showed, when the engine was stopped, the petrol in the carburettors was heated through the inlet manifold. Heat conduction from the engine and hot gases from an open inlet valve were the cause. Consider fitting 10mm insulating blocks between the carburettors and inlet manifold. Thermal images showed the choke levers underneath the carburettors were hot. Sources of heat are not what would be expected. The tests showed the carburettor float bowls did not get hot, even though they were positioned just above the exhaust manifold.

An alternative approach is to insulate the hot parts of the engine such as the exhaust manifold and downpipe. These run very hot and cool down quickly after the engine has stopped. The insulation does not have to contain the heat for very long. This may provide a more effective solution than insulating the fuel system components.

Before fitting any insulation or heat shields try to identify the hot spots. Use an infrared thermometer or test meter with a thermocouple to investigate which parts of the fuel system are getting hot. It is not always obvious where the problems lie (see Chapter 9).

Tune the engine to reduce the degree and effect of cyclic variability

A high degree of cyclic variability causes the cylinder head and exhaust system in particular to become overly hot. When driving at normal speeds this does not pose a problem. Air flowing through the engine bay and petrol through the carburettors keep things cool. When the engine is stopped this heat 'soaks' out, increasing engine bay temperature and causing the petrol to boil.

Tuning the engine to reduce the degree of cyclic variability helps to protect it from damage. It also lessens the severity of the Hot Restart Problem.

With variable jet carburettors, check the fuel level in the jet is set correctly. Modern floats or 'stay up' floats may be too light, causing the fuel level to be too low. If they are fitted, check that the springs in the suction chambers are the correct colour for your car. For example, most MGs should have red springs, the TF uses light blue and the V8 uses yellow (see Chapter 11).

If you are considering gas flowing your inlet manifold and cylinder head, think about what you want to achieve. For normal road use, this could reduce the turbulence and mixing of the petrol and air. It could make the degree of cyclic variability worse when the vehicle is used on the public highway.

Advancing the ignition timing is the most effective way of reducing exhaust temperatures. Measure your advance curve, plot it at https://classicenginesmodernfuel.org.uk/ignitionadvance/ and check it against those for your model of vehicle. (See the website for instructions). If your vehicle is fitted with a vacuum advance, check it is functioning. If your car does not have a vacuum advance, consider fitting one (Chapter 13). Consider advancing the ignition timing a few degrees (eg 5°-10°) beyond that normally recommended (Chapter 12).

DO NOT over advance the ignition timing to the point where the engine starts to pink!

Electronic ignition systems or other devices that improve the spark appear to give little benefits.

Finally, consider a session on a rolling road to ensure that your engine is optimally tuned and running as efficiently as possible.

Summary

Modern petrol and classic cars don't go together. The Manchester XPAG tests have helped to understand the cause of these problems and suggest ways they can be avoided. Hopefully, implementing such suggestions will enable owners to better enjoy their classic motoring.

16 Appendix: Tuning SU Carburettors

For some people, SU carburettors are complex parts that should not be tampered with. They believe they are difficult to setup, especially twin carburettors, and that work on them should be left to the expert. This is not true. With care and patience any person with a degree of mechanical understanding can rebuild and tune these carburettors. This appendix provides step-by-step advice on rebuilding and tuning twin SU carburettors.

The photographs in this appendix show the HS2 1¼in semi downdraft carburettors fitted to the early MG T types. While the details of newer SU and Stromberg carburettors differ, the steps to rebuild and tune them are the same.

The SU carburettor is very simple to rebuild and tune. Contrary to popular advertisements you do not need tuning plugs, balancing meters and a host of other, specialist tools. All that is needed is patience, care and an attention to detail. There are some of pitfalls for the enthusiastic amateur but these are described in this appendix.

The SU carburettor

Figures A-01 and A-02 show the main parts of the SU carburettor.

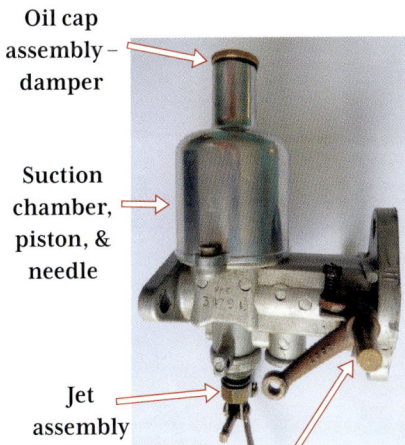

Oil cap assembly – damper

Suction chamber, piston, & needle

Jet assembly

Accelerator spindle & butterfly

A-01: Carburettor body.

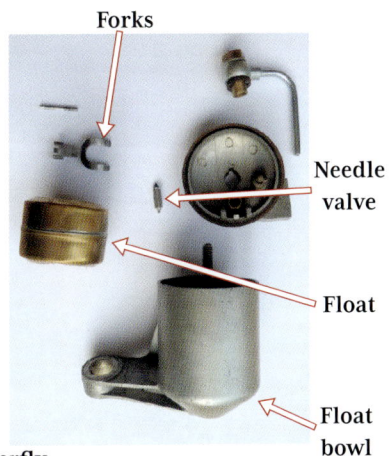

Forks

Needle valve

Float

Float bowl

A-02: Float chamber.

Before you start

Cleanliness is vital; you do not want any dirt or grit either in the carburettor or to get into the cylinders. Before you do anything, clean the area around the inlet manifold and carburettors. A product such as Gunk in combination with a stiff brush works well. When finished, wash it off with clean water and let everything dry before continuing.

Before removing anything, check if there are any air leaks in the inlet manifold or throttle spindles. When an engine is running at part throttle, there is a large pressure difference between atmospheric and that in the inlet manifold. Should there be any leaks, air will be sucked in making the mixture weaker. This will make it much harder to tune the engine.

The easiest way to test for leaks is to use a clutch and brake cleaner spray. You need to check the throttle spindles where they enter the carburettor and manifold joints.

Start the engine, and using the throttle adjusting screws (Figure A-1) increase the engine speed to about 1500rpm. Give each part a light spray with the clutch and brake cleaner. Waiting a few seconds between each one.

The spray is HIGHLY FLAMMABLE. Be careful around the exhaust manifold. Test any joints here before it gets hot.

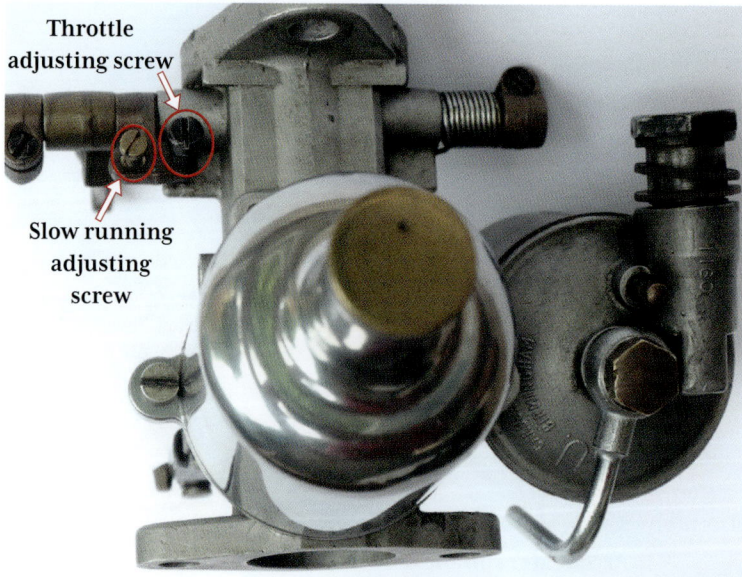

Throttle adjusting screw

Slow running adjusting screw

A-1: Throttle adjusting screw.

Should the engine note change, there is an air leak.

Air leaks in the throttle spindles is covered in the section about rebuilding the carburettors. Leaks in the inlet manifold should be fixed by replacing the gasket and using a good quality gasket and appropriate sealant.

Purchase rebuild kit(s) for your carburettors by searching 'SU rebuild kit' online. These are worth the modest expense if you are going to strip down your carburettors.

Finally, when removing the carburettors, use two plastic butter tubs or similar. Store the bolts, fastenings, etc for each carburettor separately. This will make it a lot easier when it comes to refitting them.

Removing the carburettors

While removing the carburettors is not a difficult job, it can be fiddly. It is worth searching YouTube to see if you can find a video of how to remove the carburettors from your model of car.

If your car is fitted with twin carburettors, strip down and rebuild one carburettor at a time. Take great care to note where all the parts come from.

The most complicated part to dismantle and reassemble is the jet (Figure A-2). This has a number of small seals, brass collets and copper washers – keep

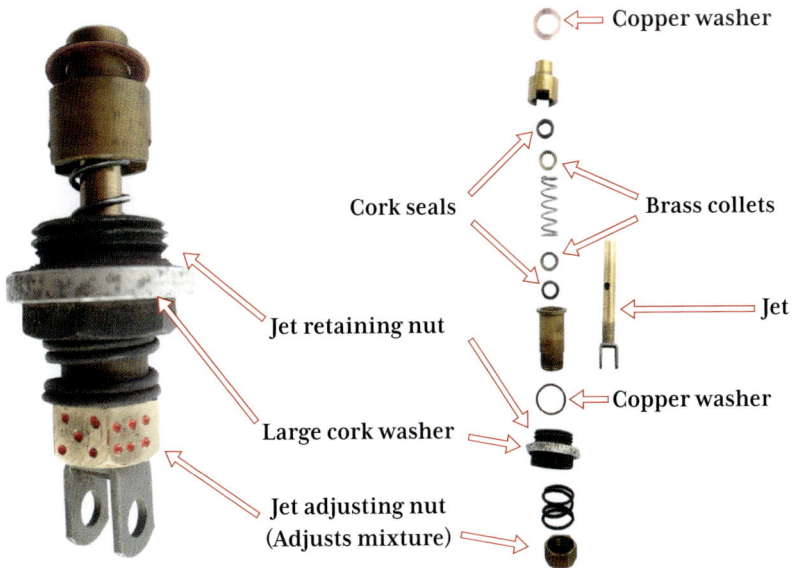

Copper washer

Cork seals

Brass collets

Jet

Jet retaining nut

Copper washer

Large cork washer

Jet adjusting nut (Adjusts mixture)

A-2: Jet assembly.

these in the correct order. If you do get in a muddle – you have ONE more chance with your second carburettor. The flat side of the brass collet goes towards the spring, the side with the indentation towards the seal.

The jet assembly is one part where later SU and Stromberg carburettors differ in detail from Figure A-2. However, the order and purpose of the parts is similar. For these carburettors, carefully strip the jet assembly, laying out the parts in order. Watch out for small thin washers such as the copper washer shown in the photograph. The easiest way to put the jet assembly back together is to start with the jet and thread all the parts onto it.

One helpful suggestion is to mark the faces of the jet-adjusting nuts on earlier SU carburettors. This makes it easier to keep the carburettors in synchronisation when adjusting the mixture. Carefully use a small drill or burr to cut one to six indentations on each face (see Figure A-2). Once painted, they give an effective way of knowing how many flats the jet-adjusting nut has been screwed down. When assembling the jets, ensure the number one faces outwards with the jet-adjusting nut fully screwed up. The mixture is made richer by screwing the nut down, weaker by screwing it up.

On the later HIF type SU carburettors the adjusting nut was replaced by a screw on the carburettor body. Turn the screw clockwise to make the mixture richer, anti-clockwise to weaken it.

On the later Stromberg carburettors, rather than changing the height of the jet, the position of the needle in the suction piston is altered. This is done using a hex allen key accessed through the top of the suction chamber. Turn the allen key clockwise to make the mixture richer, anti-clockwise to weaken it. Use the special tool to make adjustments and take care. Hold the outer tube of the tool firmly; otherwise the suction piston can turn and tear the rubber diaphragm. Note: with these carburettors, it is worth replacing the O-ring seal that stops the damper oil entering the carburettor.

The adjustments below refer to flats on the jet-adjusting nut. These translate into turns of the jet-adjusting screw or allen key. One flat is the approximately the same as $\frac{1}{6}$ of a turn of the screw or key.

With twin carburettors, treat them as identical twins – what you do to one, do to the second; what you replace on one, replace on the second.

Thoroughly clean all the parts using paraffin or white spirit and a stiff brush. Blow out all the holes, especially the one at the bottom of the float chamber that connects it to the main body of the carburettor or the jet tube in later models. These can get 'gummed up' with debris from your tank.

As an aside, while you are maintaining the fuel system it is worth cleaning any filters between the fuel tank and carburettors. On the MG TC there are TWO gauze filters, one on the bottom of the petrol tank and the other on the bottom of the fuel pump.

Items that may need replacing

• Butterfly shafts and 'bushes' – these wear, allowing air into the inlet manifold, upsetting the mixture. On the original carburettors the 'bushes' consisted of a hole in the aluminium casting. These can be badly worn. Some kits contain new shafts and PTFE bushes. Fitting these can be difficult despite what it says in the instructions. It is recommended you have the bushes fitted by a local engineering workshop especially if any machining is needed. When fitted, the new shafts should be a snug fit but still turn easily.

• Float chamber valve and seat – there are new ones in the kit. The choice of needle valves is discussed below.

• Float – floats can deteriorate over the years and start to leak. You can check them by immersing them in a pan of hot water and watching for bubbles. If you do see any, buy new floats. Beware: read the whole of this article before fitting them!

• Jet – there is a new one of these in the kit.

• Needle – This can be both worn and bent. To check it's straight, remove it from the piston and put it on a piece of glass or a flat surface. Roll it backwards and forwards holding the boss (Figure A-3). If the tip moves up and down, it is bent – replace it. On SU carburettors replace the needle with the shoulder level with the bottom of the piston. To check for wear, run your fingernail up and down the tapered part. You should be able to feel a slight roughness caused by the turning marks. If in doubt replace both needles. On my TC I use standard manufacturer-fitted ES needles.

A-3: Checking the needle is not bent.

A-4: Suction piston with a steel insert.

A-5: Suction piston with a spring.

- In prewar or immediate postwar cars, the suction pistons were made of bronze or brass. Later cars had aluminium pistons with a steel insert (Figure A-4). From around the 1950s onwards, the pistons were aluminium with a coloured spring to increase the effective weight (Figure A-5). This is discussed in more detail in Chapter 11. For the 1¼in carburettors, the brass and aluminium pistons with a steel insert weigh 238g (8.4oz) with the needle but no damper or oil. The aluminium pistons with the spring compressed should give the same effective weight.
- Check the suction piston/suction chamber are not worn or that the piston is sticking. The easiest way to do this is:
  - Invert the suction chamber and fit the piston.
  - Put your finger over the hole at the bottom/rear of the suction piston (Figure A-6).
  - Lift the suction piston and watch how quickly the suction chamber falls.
  - The suction chamber should fall smoothly and take about five seconds to fall off (Figure A-7).

A-6: Checking the suction piston.

A-7: Lifting the suction piston.

Needle valves

The needle valve in the float chamber is very important. It must not stick or leak otherwise it can flood, causing the carburettor to deliver an overly rich mixture. They are prone to wear. Figure A-8 shows the wear ring around the top of an old needle.

If there is any doubt, replace needle valves and seats. There is a wide choice. Many people recommend Grose jets (search 'grose jet float needle and seat' online) but opinions on the internet vary.

A-8: Worn needle.

Figure A-9 shows a sample of needle valves. They fall into two categories:
- Tipped or untipped – tipped valves have a rubber or Teflon tip. These give a better seal but may be more prone to wear. if you choose to use tipped needle valves, ensure that they are resistant to ethanol.
- Sprung or unsprung – Sprung needles are resistant to resonances or

vibrations in the float chamber causing leakage. The disadvantage is they do not provide a 'hard stop' and may allow the float to move upwards and cause flooding. The two sprung needles below use a pin. The Grose jets use a sprung ball. This does not move as far as the pins, and also makes them less susceptible to jamming on the forks (see below).

Steel original	Aluminium sprung and tipped	Grose jet	Brass unsprung and tipped	Nylon sprung

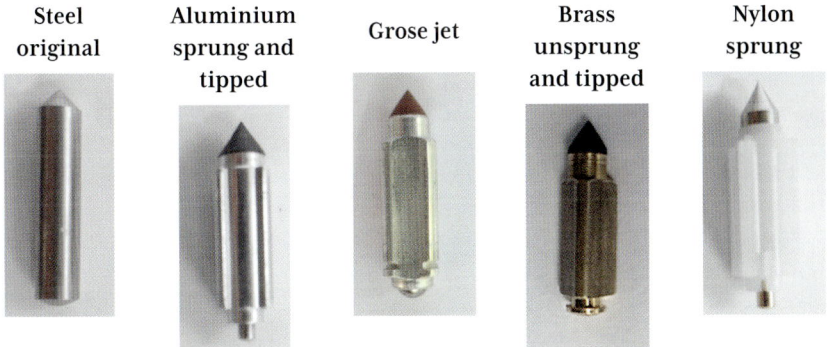

A-9: Choice of needle valves.

Reassembling

When reassembling, give moving parts a light spray of oil and check they all move freely. After spraying, wipe around the inside of the suction chamber and outside of the suction piston. These should be virtually oil free or they may stick.

Check the butterfly valve opens and closes fully under the force of the return spring. Loosen the tickover screw, and hold the carburettor body to the light to check the butterfly valve is

A-10: Checking the butterfly.

A-11: Butterfly screws.

fully closed. If not, make sure the circular disk is fitted correctly. The bevelled sides should be parallel to the carburettor body when the valve is closed. Slightly loosen the two screws, open then push the butterfly closed; this will re-seat it. Re-tighten the screws.

Ensure the split in the screws is opened and the two parts of the screw are not likely to break off. Try to set the splits to align with the airflow. This will reduce the area they present to the inflowing air.

Check you can move the jet-adjusting nut by hand. If not, be careful using a spanner – some people have 'popped' the bottom off these nuts by over tightening them! Make sure the jet can be pushed in and sits firmly against the jet-adjusting nut.

Before fully tightening the large jet-retaining nut, centre the jet. This is very important! With the jet-retaining nut loose and the carburettor body horizontal:

- Lift the suction piston to the top of its travel.
- Screw up the jet-adjusting nut (or adjusting anti-clockwise screw on later carburettors).
- Let the suction piston fall.
- Lift the suction piston again and tighten the jet-retaining nut $\frac{1}{6}$ turn.
- Let the suction piston fall.
- Repeat the last two steps until the jet-retaining nut is tight.

When the jet-retaining nut is tight, the suction piston should fall freely with a resounding 'clunk' at the bottom of its travel.

A-12: Checking the suction piston

A recommended regular maintenance check is to unscrew the oil damper. With care it can be moved to one side and used to lift the suction piston. Check the suction piston moves up and down freely, and when released at the top of its travel it falls back with a 'clunk'. If it does not, it could be that the jet needs to be re-centred or the suction piston and chamber need to be cleaned.

Hold the lid of the float chamber horizontal and check the needle valve drops freely. Ensure the forks do not fall too far as to cause it to jam against the jet when you try to raise them. The small pin on the sprung jets makes the forks more susceptible to jamming. In Figure A-13, the forks have almost dropped too far. Any further and they would jam. Should this occur, the stop on the forks

A-13: Float chamber lid.

(Figure A-14) can be bent to restrict how far they fall. (Note: If the forks do not have a tab, two small slots can be cut into them to form a tab as shown in Figure A-14)

Finally, check the fork is central and will not bind in the stud that holds the lid down.

Bend this tab to stop the forks dropping too far

A-14: Stop on the forks.

A-15: Adjusting the forks.

Before fitting the lid to the float chamber, check the setting of the forks. Place a bar across the float chamber lid and bend the forks until they just touch it as the needle valve closes. With the HS type float chamber use a $5/16$in (8mm) rod (shown in Figure A-15). With a hinged nylon float use a $1/8$in (3mm) rod. You will probably have to readjust the setting when the carburettors are fitted to the car. This is covered later.

When fitting the float chambers, set them at right angles to the body of the carburettor. Before finally tightening the retaining bolt, check you can remove the float chamber lid and float. These can foul on the screw lug for the suction chamber making them difficult to remove. If they do foul the lug, move the float chamber slightly forward (towards the engine).

Don't tighten down the float chamber lids yet. You will need to remove them to check the petrol height in the jet.

For carburettors fitted with a jet tube linking the jet to the float chamber, compress the spring, fit the brass nut, steel washer and O-ring onto the tube in that order. While holding the spring compressed, fit the pipe fully home in the float chamber and screw in the brass nut before releasing the spring.

The carburettors can now be refitted into the car. Use a gasket and good quality gasket sealant between the carburettor, spacers if fitted, and inlet manifold. It is very important these seals are airtight.

At this stage:

- *Do not* fit any air filters or the air inlet manifold. You need to have access to the suction pistons on the carburettors.
- *Do not* tighten the clamp bolt linking the accelerator rod between the two carburettors. Leave it loose. This causes a problem later when you come to start the engine – just bear that in mind!
- *Do not* fit the rod linking the choke levers choke cable. Again, this may cause a problem with you come to start the engine.

Tuning the carburettors

Before you start you will need the following 'special' tools:

1. A piece of stiff card approximately 2x4in (5x10cm).
2. 6in of ¼in (15cm of 1cm) plastic tubing. An alternative is a 'fluid transfer tool' PM-3150 from Agriemach Ltd (www.agriemach.com).
3. A small flat-bladed screwdriver.
4. Cheap set of vernier callipers (search 'cheap vernier callipers' online).

Two additional tools are useful for setting the height of the petrol in the jet (as shown in Figure A-16):

5. A spacer to place between the jet and jet-adjusting nut. This is used to drop the jet an exact amount to make it easier to see the height of the petrol. Its thickness should be equal to the distance from the top of the jet to the required petrol level. (Eg, for a petrol level ³⁄₁₆in (5mm) below the jet, this should be ³⁄₁₆in (5mm) thick).
6. Set of 'tweezers' made from piece of thin metal strip that can be used to lift the float out of the float chamber.

Steps for tuning

1. Set the petrol level in the jet by adjusting the forks in the float chamber.
2. Balance the airflow so it is the same through each carburettor.
3. Balance the mixture setting so it is the same for each carburettor.

4. Adjust the mixture.

Steps two to four will need to be repeated a small number of times to fine-tune the adjustments.

Setting the petrol level in the jet

Setting the forks in the float chambers, as described above, may not give the required petrol level in the jet. The actual height should be checked. In practice, it is very difficult to see the level of the petrol in the jet itself just by looking into the top of the carburettor. The easiest way is to use the choke lever to drop the jet. Do this slowly, watching for the petrol to appear at the top of the jet. This is where the spacer (special tool five) is useful. It allows the jet to be dropped a predefined distance. (Note: if you use a spacer, ensure the jet nuts are fully screwed in). Figure A-17 shows the jet withdrawn by ³⁄₁₆in (about 5mm).

The first step is to check the vehicle's handbook to determine if the petrol height is specified. Normally this is measured from the bridge in the carburettor. The factory handbook for the MG TC states the petrol level should be between ⅛in (3mm) and ⁵⁄₁₆in (8mm) below the bridge. A reasonable average to aim for is about half way between these figures, ie ³⁄₁₆in (5mm). Care is needed as setting the petrol level too high risks flooding.

A more conservative figure for a petrol level of ⅜in (about 10mm) below the jet is normally specified for SU carburettors.

Set the petrol height for each

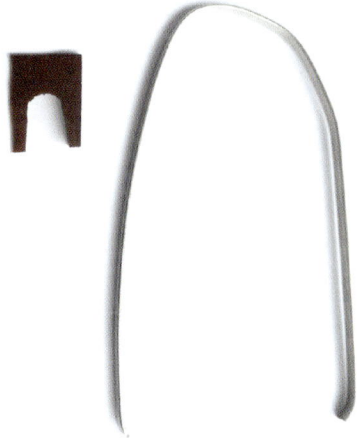
A-16: Additional 'special' tools.

A-17: Jet dropped by ³⁄₁₆in.

carburettor separately. Unscrew one of the suction chambers and remove the piston. Be careful: if there is any oil in the damper, this can end up everywhere and worse still, it is very easy to bend a needle.

It is necessary to fill the float chambers with petrol to the point where the needle valve closes off the flow. For vehicles with electric petrol pumps this is easy. Switch on the ignition and wait until the pump stops clicking, then switch off the ignition.

Note: if the electric pump does not stop ticking, there may be a petrol leak. It could be the diaphragm in the petrol pump, one of the pump valves, a petrol line or the needle valve in the float chamber. It is best to resolve this before continuing.

Before continuing, it is important to check for any petrol seepage. Check around the large jet-retaining nut, the bolt that holds the float chamber and unions connecting the fuel pipes to the carburettors. Indeed check around any part of the carburettor that contains fuel. One way to do this is to hold a clean piece of kitchen towel around the fitting for a few minutes. After removing inspect it for any dampness. Tighten the bolts holding any leaking parts. *Do not overtighten as you may strip the threads.* If this does not fix the leak, remove the bolt and clean and inspect the mating faces and any sealing washers.

Filling the float chambers and testing for petrol leaks is harder with vehicles that have mechanical petrol pumps. At this stage, it is not possible to start the engine. One possible solution is to remove the sparkplugs to reduce the load on the battery and starter and crank the engine until the float chambers are full. Some cars with mechanical fuel pumps have a lever that allows the pump to be operated manually. Check the owner's manual or the internet for your model of car.

Look into the jet and to see where the top of the petrol is. In practice this is not easy. The difficulty arises because the surface tension of the petrol in the jet 'depresses' the level as the jet is pulled down. As the jet is withdrawn, the petrol can suddenly overtop the jet by $\frac{1}{16}$in-$\frac{1}{8}$in (1-3mm). Gently blowing on the top of the jet helps the petrol settle more quickly.

The depth gauge (tool number four) can be used to measure the depression of the jet below the bridge. As an alternative, the spacer (tool number five) can be slipped between the jet and jet-adjusting nut to fix the depression of the jet.

If the petrol level is too high

If the petrol level is too high, use the measured depression of the jet to assess approximately how much too high it is. Remove the float chamber lid, bend the fork downwards (away from the top of the float chamber) by that amount. Use tool number six (above) to remove the float and tool number two to extract some petrol. Needless to say, *take extreme care* working with petrol; it is highly flammable.

Refit the float chamber lid. Remember the special washer between the top of the lid and the breather tube (Figure A-18). *Be careful,* it is very easy to over tighten and break the retaining stud off the bottom of the float chamber. Refill the float chamber and re-check the petrol height in the jet

A-18: Special washer.

If the petrol is too low

Having a petrol level that is too low is the more likely scenario. It can be difficult to address. Again, assess how much too low the petrol is. If it is less than around ¹⁄₁₆in (2mm) it can be corrected by bending the forks upwards (towards the lid of the float chamber). However, if it is greater than this, bending the forks may cause the float to foul the float chamber lid and cause flooding. This can be checked by putting a rod through the top of the float chamber and fitting the float onto the rod (Figure A-19).

The safest way to raise the petrol height is to increase the weight of the float. This can be done by adding about 0.1-0.25oz (4-8g) of solder to the bottom of the float. This corresponds to a length of approximately 1-2in (25-50mm) of 11-gauge plumber's solder. Modern digital kitchen scales are sufficiently accurate to measure the weights of the float and piece of solder.

To 'fit' the solder remove the float, coil the solder and place around the central hole on the bottom of the float. Add some flux and gently warm with a blowlamp or powerful soldering iron until it *just* melts. Avoid too much heat otherwise you risk your float falling apart.

Plastic and stay-up floats present more of a problem. It is possible to

A-19: Float fouling chamber lid.

use self-tapping screws to increase the weight of these, but ensure they do not leak afterwards. Epoxy glue is another possibility but ensure any glue you may use is petrol proof!

Re-measure the petrol height in the jet. If it is too high, solder can be scraped off the float using a sharp knife or screws removed.

Remember

You need to set the petrol level for each carburettor. Try to ensure they are as similar as possible. Compare the forks – they should be bent to the same degree; the floats should have the same weight, etc.

Balance the airflow

This is a relatively simple job that puts off many people trying to tune twin SU carburettors. While an airflow meter is ideal, tool number one, a piece of card, is quite adequate. The aim of the exercise is to set the butterfly valves on each carburettor so the same volume of air is flowing through each carburettor.

Fully screw up both the jet-adjusting nuts. If you have marked them as suggested, the number one dot should be facing you, if not add a dab of paint to that face of the nut. Screw the nuts down two complete turns. For carburettors fitted with jet-adjusting screws, these should be screwed clockwise two complete turns.

If there is one fitted, unscrew the slow running lever stop screw sufficiently to ensure it is ineffective. Screw in the throttle adjusting screws until they just start to open the butterfly.

Start the engine. This is easier said than done, especially with twin (or more carburettors) and with each one working independently. It may be prudent to temporarily reconnect the throttle linkage and use a wire or similar to operate the choke. Once the engine has started, screw in each throttle-adjusting screw to increase the tickover speed to about 1000rpm. Leave the engine running until it gets hot.

Once the engine is hot, ensure the throttle linkage between the carburettors is loose. Set the throttle stops until it is ticking over at about 500 to 750rpm. This tickover speed is not critical and it can be increased if the engine is barely running.

Place the card (tool one) horizontally over the bottom of the carburettor air inlet (Figure A-20). Move it slowly upwards restricting the airflow, and note the point the engine starts to falter or stalls. You may also notice the piston start to rise just before the engine falters. Repeat with the second carburettor.

Adjust one or both of the throttle stop screws until both carburettors exhibit the same behaviour. At this point, if you look into the air inlets, both the pistons should be floating at the same height. Also if you have a good ear, the hiss from each carburettor should be the same. (Note: you can also use the piece of plastic tubing held to the ear to hear more clearly.)

You have now balanced the airflow and can re-tighten the clamps that link the throttle spindles. From now on, it is important you make exactly the same adjustment to each throttle stop.

Reset the throttle stops to give an appropriate tickover speed (about 750rpm) (turning each by the same amount!). If one is fitted, reset the slow running control stop such that the end of the lever can be lifted by about $\frac{1}{8}$in (3mm) before it starts to open the throttle.

A-20: Measuring the airflow.

On later cars, the slow running is operated by a cam connected to the choke. There is an adjusting screw that runs on the cam to set the slow running. Again, this should be set so it is just clear of the cam with the choke fully closed. If the engine revs on a cold engine are too high or it stalls when running on the choke, the speed can be altered using this screw.

Balance the mixture

With the engine ticking over and the dampers removed, use tool four, the fine bladed screwdriver, to slowly lift one of the pistons about $\frac{1}{4}$in (6mm). The engine should falter but not stall. Repeat with the second carburettor.

If the engine stalls, that carburettor is running rich relative to the first. Screw the jet in (upwards) by $\frac{1}{6}$ turn (one flat) (for carburettors with screws, anti-clockwise by $\frac{1}{6}$ turn). Lift the piston again.

If there is no change in note, that carburettor is running weak relative to the first. Screw the jet out (downwards) by one flat and repeat (for carburettors with screws, clockwise by $\frac{1}{6}$ turn).

The aim is to get each carburettor to behave in the same way rather than to achieve an absolute result. For example, if they both stall the engine when lifted the same distance this is OK. Also, while you are doing this, keep an eye on how high each of the pistons is floating: they should be the same. If not go back to stage two and re-balance the airflow.

Once you have the mixture balanced, the same adjustment should be applied to each jet-adjusting nut. This is where the marks on the jet-adjusting nuts are useful. At this stage, check each of the jet-adjusting nuts is screwed down the same number of flats. A difference of one to two flats is acceptable.

Setting the mixture

There are two steps in setting the mixture.

The first can be difficult as the engine never behaves as expected. It is also very difficult to repeatedly lift the pistons on a running engine by the required amount using a screwdriver. Rather than trying to 'lever' the piston, it is easier to insert a flat-bladed screw driver and twist it. Use the blade to lift the piston and note how far the screwdriver rotates. It is important to follow this step to get the mixture approximately correct.

Reset the tickover to approximately 750rpm. At this stage the engine should be running relatively smoothly. Gently lift the piston of one carburettor by $\frac{1}{32}$in to $\frac{1}{16}$in (1-2mm) and hold it. The engine speed should increase slightly and then settle back down.

A-21: Lifting the suction piston.

If the engine stalls, the mixture is too weak: screw out BOTH jet-adjusting nuts by one flat (jet-adjusting screws clockwise by $\frac{1}{6}$ of a turn). If the engine speeds up and continues to rev, the mixture is too rich: screw up BOTH jet-adjusting nuts by one flat (jet-adjusting screws anti-clockwise by $\frac{1}{6}$ turn).

Repeat the procedure on the second carburettor. You may find that you need to make different adjustments to each carburettor. This is OK as long as they do not differ by more than two flats of the jet-adjusting nut. You may find it necessary to repeat stage three.

When you believe the mixture is properly set, rev the engine and watch that the suction pistons in the carburettors float to the same height and rise and fall at the same rate. With the engine ticking-over, the exhaust should be emitting a regular 'poph ... poph ... poph' sound.

Replace the oil dampers and check the piston is 'hard' to push upwards, but still falls back down freely.

Final tuning

You can now refit the jet-coupling rod, ensuring the holes match those in the jets. You may have to adjust its length to fit. Once fitted, pull out the choke, push it back in again and check both jets are fully against the jet-stop nuts. If not, investigate the cause.

Refit the air inlet manifold, air filters, etc. You are now ready for the final stage of tuning; this is more accurate and, more importantly, more fun.

Unfortunately, modern petrol makes this method less sensitive than it used to be. However, it is still a good way of ensuring the mixture is correct.

Take the car out for a 15 to 20-mile drive. "Just checking it's running OK" is a good excuse! Try to avoid idling the engine, especially when you get back home.

As soon as the engine has cooled sufficiently, remove the sparkplugs, being careful to keep them in order. Look at the colour of the electrodes: cylinders one and two should be the same colour, as should three and four, and the colour of the insulation around the electrodes on all the sparkplugs should be the same. The colour depends on the brand and grade of petrol you are using. In some cases they will colour brown, grey is more probable, possibly even red.

If they are:
• Dark brown, dark grey, dark red, black or sooty – the mixture is too rich. Screw both jet-adjusting nuts in (upwards) (anti-clockwise for screws). Two to three flats if the plugs are sooty, one to two flats if they are a dark colour.

A-23: A grey plug colour.

A-22: The ideal plug colour.

- Fawn (the same colour as a digestive biscuit, the colour produced by classic petrol). With modern petrol this may be a dark or silvery grey when the mixture is correct.
- Light fawn, light grey or white – the mixture is too weak. Screw both jet-adjusting nuts out (downwards) (clockwise for screws). Two to three flats if the plugs are white, one to two flats if they are a light colour.

When adjusting the jet-adjusting nuts, move them by NO MORE than three flats or ½ turn at a time. If the front two plugs are a different colour from the rear two, correct this by adjusting only one jet-adjusting nut. Changing one carburettor by more than two flats from the other risks upsetting the balance. It may be necessary to re-balance the airflow.

This method has the advantage of setting the mixture to match your 'normal' driving conditions, allows for the effect of the air filter, is easier, more reliable and definitely more fun. It is also worth checking the colour of the sparkplugs as part of your regular maintenance routine.

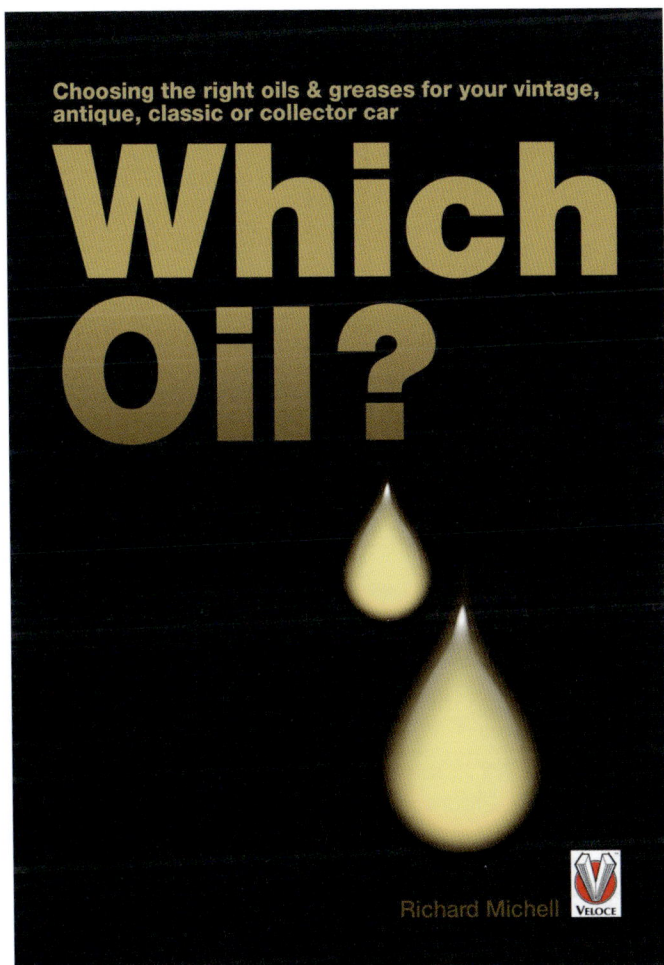

Choosing the right oils & greases for your vintage, antique, classic or collector car

Which Oil?

Richard Michell

VELOCE

ISBN: 978-1-845843-65-6

Paperback • 19.5x13.9cm • 128 pages

36 b&w pictures and line drawings

For more information and price details, visit our website at www.veloce.co.uk

email: info@veloce.co.uk · Tel: +44(0)1305 260068

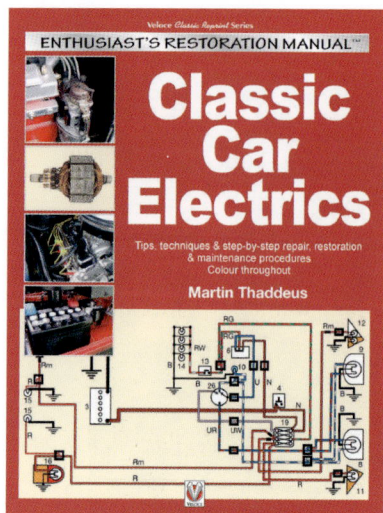

ISBN: 978-1-787111-01-1

Paperback • 27x20.7cm • 96 pages

301 pictures

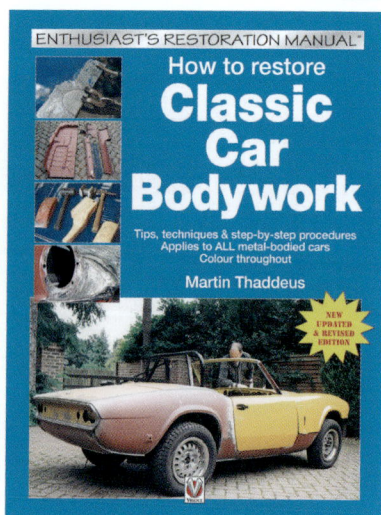

ISBN: 978-1-787111-67-7

Paperback • 27x20.7cm • 128 pages

350 pictures

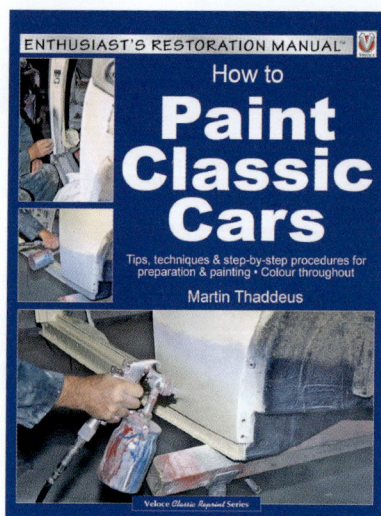

ISBN: 978-1-787111-42-4

Paperback • 27x20.7cm • 96 pages

213 pictures

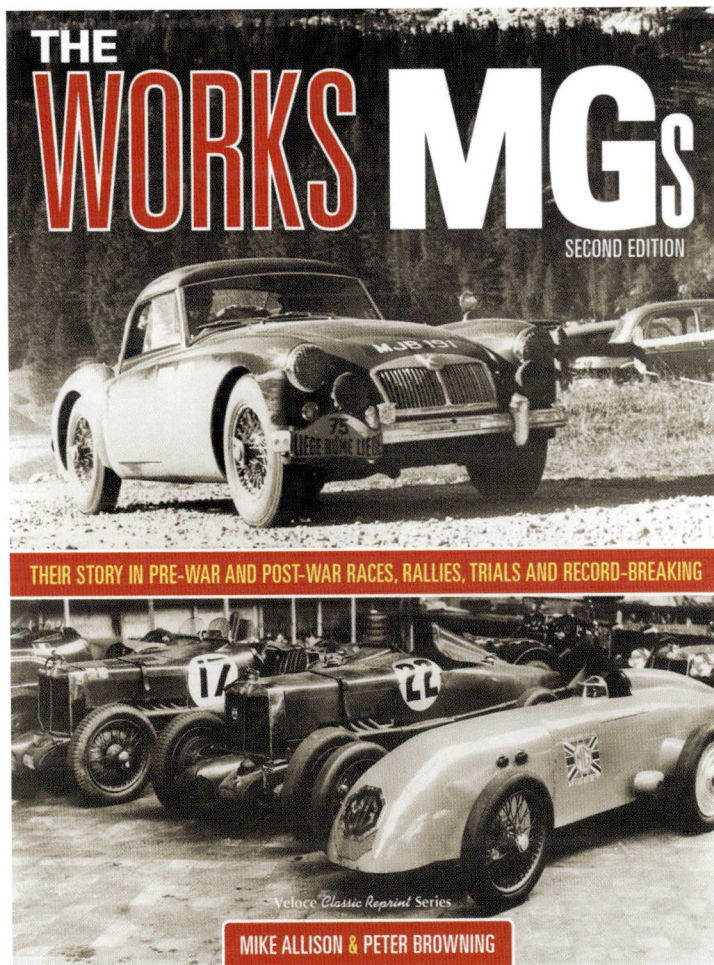

Index